# DAIRY MICROBIOLOGY

*Volume 1*

The Microbiology of Milk

*Edited by*

R. K. ROBINSON
M.A., D. Phil.

*Department of Food Science,*
*University of Reading, UK*

APPLIED SCIENCE PUBLISHERS
LONDON and NEW JERSEY

APPLIED SCIENCE PUBLISHERS LTD
Ripple Road, Barking, Essex, England
APPLIED SCIENCE PUBLISHERS, INC.
Englewood, New Jersey 07631, USA

**British Library Cataloguing in Publication Data**

Dairy microbiology.
Vol. 1: The microbiology of milk
1. Dairy products 2. Food microbiology
I. Robinson, R. K.
630.2′76 QR121

ISBN 0-85334-948-7

WITH 31 TABLES AND 37 ILLUSTRATIONS


Printed in Great Britain by Galliard (Printers) Ltd, Great Yarmouth

# DAIRY MICROBIOLOGY

*Volume 1*

The Microbiology of Milk

# Preface

The chemical composition of milk renders its inclusion in the human diet extremely valuable, and yet it is these same components that support the myriad of bacteria, yeasts and moulds that may be associated with dairy products. The activity of some of these micro-organisms is clearly advantageous, a view confirmed by the numerous varieties of cheese on the consumer market, but equally evident is the fact that uncontrolled microbial activity can lead to spoilage and, in some cases, the unwelcome development of pathogens. This susceptibility of milk and milk products to deterioration, together with the resultant loss of consumer appeal, have stimulated considerable interest in dairy microbiology. As a consequence, an extensive body of information is available concerning the behaviour of micro-organisms in milk and its derivatives, and it is now widely accepted that, with proper handling, the contamination of milk, or at least any subsequent microbial activity, can be contained within acceptable limits.

In many countries, this view has been crystallised in the form of regulations covering milk production and distribution, and/or the derivation of legal standards covering the maximum levels of contamination permitted in consumer items. Alternatively, standard methods of examination have been evolved, and these are employed to support a monitoring system that is basically advisory in nature. However, irrespective of the chosen approach, the outcome has been a gradual elevation of the microbiological quality of dairy products, so that at the present time, the industry enjoys an enviable reputation in terms of the hygienic quality of its output.

However, the diverse nature of dairy products ranging from bottled milk through to highly processed consumer items, means that not only does current industrial and regulatory practice vary considerably, but also that the available information is widely disseminated. As a consequence, any student of food or dairy science, or indeed any newcomer to the industry, faces a daunting task in attempting to acquaint him- or herself with a basic knowledge of the microbiology relevant to the commodity spectrum.

One solution to this dilemma entails the provision, under one cover, of a series of authoritative treatises that seek to highlight the microbiological aspects of milk production and processing, and it was the obvious attraction of this proposal that provided the motivation behind this present venture. The format of the book as a multi-author text became, therefore, virtually inevitable, for only in this way could the diverse aspects of the subject be handled with appropriate expertise. It was also evident that the disparate nature of dairy products would invalidate any attempt to impose a uniform or stereotyped presentation, and hence no attempt has been made to inhibit the individual inclination of authors in terms of presentation or content.

A conscious attempt has been made, however, to present dairy microbiology as a definable entity. Thus, starting with the nature of milk itself and the micro-organisms associated with it, the book seeks to cover a comprehensive range of products and processes, and although bound in two volumes, it is intended that readers should regard it as one complete unit. The validity of this approach is, of course, open to debate but, as a method of providing information for those interested in the microbiological quality of dairy products, it is probable that no other treatment would have elicited such original contributions from the various authors.

This latter point merits especial emphasis, for it is the endeavours of the individual authors that have brought these volumes into being. Their expert knowledge of particular areas of dairy microbiology has made my job as editor both enjoyable and rewarding, and my sincere gratitude is extended to all who have contributed to these two volumes.

In conclusion, the role of Applied Science Publishers Ltd must not go unrecorded, and if this text contributes in any way to our understanding of the microbiology of milk and milk products, then their enterprise will have been totally worthwhile.

R. K. ROBINSON

# Contents

# List of Contributors

W. BANKS, B.Sc., Ph.D.,

*The Hannah Research Institute, Ayr, UK.*

A. J. BRAMLEY, B.Sc., Ph.D.,

*National Institute for Research in Dairying, Shinfield, Reading, UK.*

CHRISTINA M. COUSINS, B.Sc., Ph.D.,

*National Institute for Research in Dairying, Shinfield, Reading, UK.*

D. G. DALGLEISH, B.Sc., Ph.D.,

*The Hannah Research Institute, Ayr, UK.*

A. GILMOUR, B.Sc., Ph.D.,

*Department of Agricultural and Food Bacteriology, Queen's University, Belfast, UK.*

H. R. LOVELL, M.Sc., N.D.D.T., F.I.F.S.T., F.A.I.F.S.T., M.I.Biol., M.Inst.Pkg.,

*School of Food Studies, Queensland Agricultural College, Lawes, Queensland 4543, Australia.*

PROFESSOR F. E. NELSON, B.S., M.S., Ph.D.,

*Emeritus Professor of Food Science, Department of Dairy and Food Science, University of Arizona, Tucson, Arizona 85721, USA.*

Professor J. A. F. Rook, Ph.D., D.Sc., F.R.S.C., F.I.Biol., F.R.S.E.,
*The Hannah Research Institute, Ayr, UK.*

M. T. Rowe, B.Sc.,
*Department of Agricultural and Food Bacteriology, Queen's University, Belfast, UK.*

Professor Robert R. Zall, B.Sc., M.S., Ph.D.,
*Department of Food Science, Cornell University, Stocking Hall, Ithaca, New York 14853, USA.*

# 1

# Milk and Milk Processing

W. Banks, D. G. Dalgleish and J. A. F. Rook

*The Hannah Research Institute, Ayr, UK*

Milk is secreted by mammals for the nourishment of their young and the milks of all species are complex biological fluids, containing a wide variety of different constituents and possessing unique physical characteristics. The major constituent of cow's milk is water, the remainder consists largely of lipids, proteins and carbohydrate materials synthesised within the mammary gland. Also present, but in smaller quantities, are mineral components and other water-soluble and lipid-soluble materials transferred directly from blood plasma, specific blood proteins and traces of enzymes and intermediates of mammary synthesis. Much of the lipid is in the form of small globules surrounded by a membrane that separates the fat core from the aqueous phase, and the major proteins, the caseins, are present in the form of aggregated particles known as micelles. The physical form of the lipid and the caseins affects profoundly the characteristics of whole milk, and has important consequences for milk processing.

## THE CONSTITUENTS OF MILK

### Lipids and Fatty Acids

The major lipid component of cow's milk is triglyceride (accounting for 97–98 % of the total lipid) but additionally there are small amounts of diglyceride (0·25–0·48 %), monoglyceride (0·016–0·038 %), cholesterol ester (trace), cholesterol (0·22–0·41 %), free fatty acids (0·10–0·44 %) and phospholipid (0·2–1·0 %); higher values for the proportions of the diglyceride and free fatty acids may be observed in samples that have been subject to lipolysis. The major fatty acids of the lipid contain 16 or 18 carbon (C) atoms, and a high proportion of the $C_{18}$ acids are mono-unsaturated. In common with the milks of other herbivorous species, cow's milk fat contains a high proportion of short-chain (4 to 14 C atoms) fatty acids and very small amounts of di-unsaturated fatty acids. There are, in addition, trace amounts of a large number of branched-chain saturated acids, and positional and configurational isomers of unsaturated acids. The acids present in cow's milk fat include all the normal saturated fatty acids from $C_2$ to $C_{28}$, monomethyl branched-chain fatty acids from $C_{16}$ to $C_{28}$, multimethyl branched-chain fatty acids from $C_{16}$ to $C_{28}$, *cis*- and *trans*-mono-enoic fatty acids from $C_{10}$ to $C_{26}$, numerous di- and polyenoic fatty acids, keto and hydroxy fatty acids and cyclohexyl fatty acids. Patton and Jensen (1975) reported 437 identified acids with others still to be characterised.

The distribution of fatty acids between the various positions of the glycerol moiety of triglycerides varies with the fatty acid. The short-chain fatty acids, butyric and hexanoic, are esterified almost wholly in position *sn*-3.† As the chain length of saturated fatty acids increases, a progressively higher proportion of the acid is located in positions *sn*-2 and then *sn*-1, and both $C_{16:0}$ and $C_{18:0}$ are found in greatest concentration in position *sn*-1. Unsaturated fatty acids are present in higher proportions in position *sn*-3 than in other positions, and *trans*-unsaturated acids are present almost entirely in positions *sn*-1 and *sn*-3.

The isolation of milk fat globule membrane of uniform lipid composition poses technical problems, and there is uncertainty about the amount and composition of the lipid component. Values for total lipid have ranged from 0·5–1·1 mg lipid per mg protein, due mainly to variations in the amount of triglyceride. Phospholipid values are fairly constant at about 35·5 μg lipid phosphorus per mg protein, but there is great variation in reported values for cholesterol and cholesterol ester (0·1–0·8% and 0·2–5·2%, respectively, of the total lipid). The contribution of individual phospholipids (as mole % of the total lipid phosphorus) is phosphatidylcholine, 38·0; phosphatidylethanolamine, 27·4; phosphatidylserine, 5·0; phosphatidylinositol, 8·0; sphingomyelin, 19·1 and lysophospholipid, 2·3.

Typical values for the composition of the fatty acids of the different lipid fractions of milk fat are given in Table I.

**Proteins**

The principal families of proteins in bovine milk are $\alpha_{s_1}$-caseins, $\alpha_{s_2}$-caseins, β-caseins, κ-caseins, α-lactalbumins, β-lactoglobulins, serum albumin and

† Stereospecific numbering system: in natural triglycerides the glycerol moiety has the L-configuration and the primary position occupied by a fatty acid is denoted *sn*-1, the secondary position *sn*-2 and the remaining position *sn*-3.

TABLE I

TYPICAL VALUES (MOLES PER CENT) FOR THE FATTY ACID COMPOSITION OF THE TRIGLYCERIDES, OF THE FATTY ACIDS ESTERIFIED TO THE VARIOUS POSITIONS OF THE GLYCEROL MOIETY AND OF THE CHOLESTEROL ESTERS AND INDIVIDUAL GLYCEROPHOSPHOLIPIDS OF COW'S MILK FAT[a]

| *Fatty acid*[b] | *Triglycerides* | | | | *Cholesterol*[c] *ester* | *Phosphatidyl-choline* | *Phosphatidyl-ethanolamine* | *Phosphatidyl-serine* |
|---|---|---|---|---|---|---|---|---|
| | *Total fraction* | *Positions of glycerol moiety* | | | | | | |
| | | *sn-1* | *sn-2* | *sn-3* | | | | |
| 4:0 | 11·3 (3·3)[d] | 5·0 | 2·9 | 43·3 | — | — | — | — |
| 6:0 | 4·8 (1·6) | 3·0 | 4·8 | 10·8 | — | — | — | — |
| 8:0 | 2·3 (1·3) | 0·9 | 2·3 | 2·2 | — | — | — | — |
| 10:0 | 4·2 (3·0) | 2·5 | 6·1 | 3·6 | 4·2 | — | — | — |
| 12:0 | 3·9 (3·1) | 3·1 | 6·0 | 3·5 | 5·2 | — | — | — |
| 14:0 | 11·5 (9·5) | 10·5 | 20·4 | 7·1 | 7·7 | 18·4 | 1·5 | 3·5 |
| 15:0 | (0·6) | — | — | — | 2·2 | 2·1 | 0·5 | — |
| 16:0 | 27·1 (26·3) | 35·9 | 32·8 | 10·1 | 27·0 | 36·4 | 11·7 | 8·4 |
| 16:1 | 2·0 (2·3) | 2·9 | 2·1 | 0·9 | 12·0 | 0·6 | 2·1 | — |
| 18:0 | 10·4 (14·6) | 14·7 | 6·4 | 4·0 | 6·1 | 11·1 | 10·5 | 36·0 |
| 18:1 | 21·1 (29·8) | 20·6 | 13·7 | 14·9 | 12·6 | 25·7 | 46·7 | 45·8 |
| 18:2 | 1·4 (2·4) | 1·2 | 2·5 | −0·5 | 9·3 | 5·3 | 12·4 | 7·6 |
| 18:3 | — (0·8) | — | — | — | — | 1·1 | 3·4 | 1·9 |
| 20:3 | — — | — | — | — | — | 1·0 | 1·4 | — |
| 20:4 | — — | — | — | — | — | 0·7 | 0·9 | — |

[a] From Christie (1978).
[b] Number of carbon atoms, number of double bonds.
[c] Also contains 13·1% of 13:1.
[d] Weight % of total fatty acids.

TABLE II

THE PROTEINS OF COW'S MILK

| *Family* | *Approximate molecular weight (daltons)* | *Genetic variants* | *Comments* |
|---|---|---|---|
| $\alpha_{s_1}$-caseins | 23 000 | *A, B, C, D* | Major component contains 7 and minor component 9 phosphorylated residues |
| $\alpha_{s_2}$-caseins | 25 000 | *A, D* | Components contain 10, 11, 12 or 13 phosphorylated residues |
| κ-caseins | 19 000[a] | *A, B* | Components containing 0 to 5 carbohydrate chains |
| β-caseins | 24 000 | $A^1, A^2, A^3, B, Bz, D, E, C$ | |
| γ-caseins | 20 500 | $A^1, A^2, A^3, B$ | |
| | 11 900 | $A^2, A^3, B$ | |
| | 11 600 | *A, B* | |
| β-lactoglobulin | 18 300 | *A, B, C, D, Dr* | |
| α-lactalbumin | 14 200 | *A, B* | Several minor proteins, some glycoproteins |
| Bovine serum albumin | 66 000 | | Heterogeneous |
| Immunoglobulins | | | Heterogeneous |
| IgGl | 163 000 | | |
| IgG2 | 150 000 | | |
| IgA | 390 000 | | |
| IgM | 950 000 | | |
| Proteose peptones | | | Heterogeneous |
| component 3[b] | — | | |
| component 5 | 12 500 | | |
| component 8 fast | 3 500 | | |
| Component 8 slow | — | | |
| Lactoferrin | 77 000 and 93 000 | | |

[a] Monomer molecular weight.

[b] As identified by moving-boundary electrophoresis of milk serum proteins.

immunoglobulins $IgG_1$, $IgG_2$, IgA and IgM. With the exception of the immunoglobulins which are coded for by a multigene complex, each family is coded for by a single gene. The various proteins of cow's milk are listed in Table II; not all of the recognised genetic variants are to be found in British breeds of cattle.

The caseins are milk-specific proteins. The amino acid compositions of the major caseins are given in Table III. The $\alpha_s$- and $\beta$-families are

TABLE III
AMINO ACID COMPOSITIONS OF THE MAJOR CASEINS (MOLES/MOLE PROTEIN)

| | $\alpha_{s_1}$ (*type B*) | $\alpha_{s_2}$ (*type A*) | $\beta$ (*type $A^1$*) | $\kappa$ (*type B*) |
|---|---|---|---|---|
| Lysine | 14 | 24 | 11 | 9 |
| Histidine | 5 | 3 | 6 | 3 |
| Arginine | 6 | 6 | 4 | 5 |
| Aspartic acid | 15 | 18 | 9 | 11 |
| Threonine | 5 | 15 | 9 | 14 |
| Serine | 16 | 17 | 16 | 13 |
| Glutamic acid | 39 | 40 | 39 | 27 |
| Proline | 17 | 10 | 34 | 20 |
| Glycine | 9 | 2 | 5 | 2 |
| Alanine | 9 | 8 | 5 | 15 |
| Valine | 11 | 14 | 19 | 11 |
| Methionine | 5 | 4 | 6 | 2 |
| Isoleucine | 11 | 11 | 10 | 13 |
| Leucine | 17 | 13 | 22 | 8 |
| Tyrosine | 10 | 12 | 4 | 9 |
| Phenylalanine | 8 | 6 | 9 | 4 |
| Trytophan | 2 | 2 | 1 | 1 |
| Cysteine | 0 | 2 | 0 | 2 |
| Phospherine | 8 | 11 | 5 | 1 |

characterised by serine-bound phosphates some of which are grouped together in a phosphate centre and confer special characteristics, and a high content of proline residues distributed throughout the molecule which prevents the formation of much secondary structure. The fraction originally referred to as $\alpha_s$-casein is now known to consist of two different groups of proteins the $\alpha_{s1,0}$- and $\alpha_{s_2}$-caseins. The most common genetic variant (*B*) of the predominant component, $\alpha_{s_1}$-casein, has a polypeptide chain of 199 residues, eight of which are phosphorylated serines; the *C* and *D* variants have single amino acid substitutions at positions 192 and 53, respectively, whereas in the *A* variant the peptide 14–26 is deleted. The minor component of the $\alpha_{s1,0}$-group, $\alpha_{s0}$-casein has a phosphoserine instead of a serine at position 41. The $\alpha_{s2}$-caseins have a polypeptide chain of 207 residues which, unlike $\alpha_{s1}$-caseins, includes two cysteine residues. The various $\alpha_{s2}$-caseins have 10, 11, 12 and 13 phosphorylated serine residues, and a polymorphism within the group has been reported.

The $\beta$-caseins have a polypeptide chain of 209 amino acids which include five phosphoserine residues (four in the *C* variant). The $A^2$ variant is most common, and the other variants are produced by the substitution of one ($A^1$, $A^3$, *E*), two (*B*) or three (*C*) amino acids in the $A^2$ chain. A protein fraction known as $\gamma$-casein is invariably present in bovine milk as secreted, and the $\gamma$-caseins have now been shown to be identical with fragments of the $\beta$-casein molecule and are thought to arise by post-translational proteolysis. $\gamma_1$-casein is identical with residues 29–209, $\gamma_2$ with residues 106–209 and $\gamma_3$ with residues 108–209. There are genetic variants of the $\gamma$-caseins that correspond to the known differences in the variants of $\beta$-casein. The *N*-terminal peptides produced on cleavage of the $\beta$-casein molecule are the source of some of the 'proteose-peptones' of milk.

$\kappa$-Casein consists of a polypeptide chain of 169 residues, with cysteine residues at positions 11 and 88 and a serine phosphate residue at position 149. The genetic variants *A* and *B* differ in the amino acids at positions 136 and 148. The molecule can contain up to five carbohydrate (*N*-acetylgalactosamine, galactose and *N*-acetylneuraminic acid) side chains, but the sites of glycosylation are not all known. The polypeptide chain is cleaved specifically at a phenylalanine–methionine linkage (105–106) by chymosin (rennin) and some other proteases, fragment 1–105 being termed *para*-$\kappa$-casein and 106–169 glycomacropeptide or caseinomacropeptide.

The major non-casein proteins (often referred to as whey proteins), $\alpha$-lactalbumin and $\beta$-lactoglobulin have, respectively, polypeptide chains of 123 and 162 amino acids, with the former containing four disulphide bonds and the latter two disulphide bonds and one sulphydryl group. Genetic variants arise by amino acid substitution in the main chain, with the exception of $\beta$-lactoglobulin Dr (Droughtmaster) which has the same amino acid composition as the *A* variant of $\beta$-lactoglobulin, but has a carbohydrate moiety attached.

Two families of iron-binding proteins are found in milk in trace amounts, one apparently identical with the transferrin of blood, the other characteristic of milk and certain other body secretions and designated lactoferrin.

Milk also contains a protein component identical in every respect with bovine blood serum albumin, and there is a transfer to milk of antibodies or immunoglobulins, particularly evident in the first secretion following parturition (the colostrum), which have their origin in blood plasma or the plasma cells of the mammary gland.

### Carbohydrates

The disaccharide lactose (β-D-galactopyranosyl-(1–4)-D-glucopyranose) is the predominant and distinctive carbohydrate of milk but there are, in addition, very low concentrations of monosaccharides, including glucose and galactose, neutral and acid oligosaccharides and the peptide and protein-bound carbohydrates.

### Salts

The major cations of milk are calcium, magnesium, sodium and potassium and the major anions are phosphate, chloride and citrate. Sodium, potassium and chloride are present in milk almost exclusively in a soluble, ionised form, but the other constituents are present mainly as unionised or weakly ionised salts with, apart from citrate, considerable amounts being colloidally bound to the caseinate complex.

### Minor or Trace Constituents

A large number of mineral elements are present in normal milk in trace amounts. A wide variety of non-protein nitrogenous compounds also occur, and the concentration of the major component urea, varies directly with its concentration in blood plasma. Also present in detectable amounts are various other water- and lipid-soluble substances including the vitamins and, perhaps accidentally, numerous enzymes including proteases and lipases, and intermediates of milk synthesis; citrate, present in substantial amounts, may be included in this category.

## THE COMPOSITION OF MILK

Typical values for the composition of bulk milk are given in Table IV. Composition varies widely, however, under the influence of a number of factors, of which the most important are breed, species and individuality, stage of lactation, age, disease (udder disease in particular) and nutrition. These effects have been studied most widely in relation to gross composition, but variations in the composition of minor components may be of equal or greater magnitude.

# TABLE IV

## TYPICAL VALUES FOR THE COMPOSITION OF COW'S MILK[a]

| *Proximate composition* | | *Protein composition* | | *Non-protein nitrogenous constituents* | | *Trace elements* | | *Vitamins* | |
|---|---|---|---|---|---|---|---|---|---|
| *Constituent* | *Concentration* (*g litre*$^{-1}$) | *Constituent* | *Concentration* (*g litre*$^{-1}$) | *Constituent* | *Concentration* (*mg N litre*$^{-1}$) | *Constituent* | *Concentration* ($\mu$*g litre*$^{-1}$) | *Constituent* | *Concentration* (*mg litre*$^{-1}$) |
| Fat | 37 | Casein | 26·0 | Ammonia | 6·7 | Aluminium | 150–1 000 | Water-soluble | |
| True protein | 34 | $\alpha_{s_1}$ | 11·1 | α-amino acids | 37·4 | Arsenic | 30–60 | Thiamine | 0·44 |
| (protein N × 6·38) | | $\alpha_{s_2}$ | 1·7 | Hippuric acid | 4·0 | Boron | 100–400 | Riboflavin | 1·75 |
| Non-protein N | 1·9 | $\beta$ | 8·2 | Urea | 83·3 | Cobalt | 0·2–1·4 | Nicotinic acid | 0·94 |
| Lactose | 48 | $\gamma$ | 1·2 | Creatine | 12·5 | Copper | 30–170 | Pyridoxine | 0·64 |
| Ash | 7·0 | $\kappa$ | 3·7 | Creatinine | 1·8 | Iron | c. 300 | Pantothenic | |
| Calcium | 1·25 | Other components of | | Uric acid | 7·6 | Manganese | 12–35 | acid | 3·46 |
| Phosphorus | 0·96 | the casein micelles | | Orotic acid | 12·0 | Molybdenum | 20–150 | Biotin | 0·031 |
| Magnesium | 0·12 | Calcium | 0·80 | Carnitine | 0·9 | Silicon | 870–2 270 | Folic acid | 0·002 8 |
| Sodium | 0·58 | Magnesium | 0·40 | Acetylcarnitine | 0·8 | Zinc | 1 000–6 000 | Vitamin B12 | 0·004 3 |
| Potassium | 1·38 | Phosphorus | | Phosphorylethanolamine | 8·3 | Bromine | 180–250 | Choline | 121 |
| Chloride | 1·03 | organic | 0·22 | Glycerophosphorylethanolamine | 2·9 | Iodine | 10–80 | Inositol | 110 |
| Sulphur | 0·30 | inorganic | 0·38 | *N*-acetylglucosamine | 7·0 | | | Ascorbic acid | 20 |
| Citric acid | 1·75 | Citrate (as citric acid) | 0·15 | *N*-acetylglucosamine-1-phosphate | 7·3 | | | Fat Soluble | |
| $CO_2$ | 0·20 | α-lactalbumin | 0·7 | | | | | Vitamin A | 0·45 |
| | | β-lactoglobulin | 3·0 | | | | | | |
| | | Serum albumin | 0·3 | | | | | | |
| | | Immunoglobulin Ig | 0·6 | | | | | | |
| | | Lactoferrin | >0·018 | | | | | | |
| | | | | Total | 193·0 | | | | |

[a] From Jenness (1974).

Breed has a distinctive effect on the gross composition of milk—of the dairy breeds, those of the Channel Islands give milks with an unusually high content of the major constituents, fat in particular, but, within a breed, a wide range of composition is observed between individuals within a herd. These differences are partly genetic in origin and partly the result of environmental and physiological factors.

The most profound change in milk composition is that which occurs with the progress of lactation. Colostrum, the first secretion removed from the udder at the beginning of a lactation, has a high concentration of fat and protein and a low concentration of lactose; the concentration of all the protein fractions are high, but there is an exceptionally high proportion of globulin, mainly immunoglobulin, which persists for one or two days only. As lactation continues, there is a rapid reduction in fat, and in total and individual protein contents, to minima at about 6 to 8 weeks of lactation, and an increase in lactose content to a maximum between weeks 2 and 4 of lactation. Throughout the rest of lactation these changes are reversed, slowly at first and then more rapidly. There are comparable changes in the concentrations of the major minerals and many of the minor components. Colostral fat has a lower proportion of short-chain fatty acids, especially butyric acid, and a higher proportion of palmitic acid than the milk fat removed subsequently during the first week of lactation. The proportions of short-chain fatty acids tend to increase throughout the first 8 to 10 weeks of lactation, mainly at the expense of stearic and octadecanoic acids; that of palmitic acid remains fairly constant. Subsequent lactational changes are small.

Lactational trends in composition are modified by environmental factors, the most important of which are udder disease (bacterial infection-mastitis) and the quantity and composition of diet. Mastitis, in clinical or sub-clinical form, characteristically causes a fall in lactose and potassium concentrations, and an increase in those of sodium and chloride, and of the proteins that are derived from the blood serum; there is additionally a massive entry of somatic cells which may release degradative enzymes. The contents of fat and of the specific milk proteins may also be altered, especially when clinical disease is present.

The effects of diet are complex. Fat content and composition are affected both by the amount and type of fat included in the diet, and by the composition of the non-lipid components. Additions of saturated fats to the diet may give a slight increase in milk fat content with a change of composition in the direction of that of the added fat. Highly unsaturated fats normally cause a depression in milk fat content, but any direct effects on composition are modified by an extensive hydrogenation of unsaturated fatty acids within the reticulo-rumen; a high proportion of the stearic acid taken up from plasma triglycerides by the mammary gland is, however, desaturated to oleic acid. An extreme depression in milk-fat content may be induced by a dietary addition of cod liver oil which has a high content of long-chain (20 and 22 C atoms) polyunsaturated fatty acids. Decreasing the amount or particle size of the forage in the diet and increasing the amount of soluble carbohydrates gives a similar depression, and both effects are thought to be due mainly to an alteration in the fermentation pattern within the rumen, and in the fermentation products that are absorbed. A change of fermentation pattern that gives rise to a reduced fat content usually increases the milk protein content. Changes in the amount and composition of the diet may also alter permanently the trends in composition anticipated during the remainder of lactation, and influence the partition of use of nutrients between the mammary gland and other body tissues; a factor of great importance in early lactation, when body stores of energy may be extensively mobilised to support a high level of milk secretion.

## THE PROPERTIES OF MILK

### The Proteins

The physicochemical properties of milk are determined largely by the behaviour of the casein complex. Almost all of the caseins of freshly secreted milk are contained in the aggregated particles referred to as micelles, and most of the soluble milk constituents and milk lipids interact with the caseins.

#### *The Caseins*

The caseins were defined originally as the phosphoproteins precipitated from milk at pH 4·6; the relatively low iso-electric point reflects the preponderance of acidic over basic residues and the presence of the phosphorylated seryl residues. The phosphate side groups, which titrate in the pH range 6·0–7·0, also have a major influence on the charge of $\alpha_s$- and $\beta$-caseins. They are concentrated towards the *N*-terminal end of the molecules and, as the remainder of the amino acid chain is largely hydrophobic, this gives the molecules amphiphilic properties.

*$\alpha_s$- and $\beta$-caseins.* These caseins are distinguished by their precipitation in the presence of calcium ions, but there are differences in behaviour between the $\alpha_s$- and the $\beta$-caseins and also between the $\alpha_{s_1}$- and $\alpha_{s_2}$-caseins; only

qualitative information is available for $\alpha_{s_2}$-casein. In the absence of $Ca^{2+}$, $\alpha_{s_1}$-casein tends to aggregate as the concentration is increased, but the aggregates contain not more than six units, whereas $\beta$-casein, which is more strongly hydrophobic, under similar circumstances forms aggregates of about 30 units and produces a soap-like micelle. The precipitation of $\alpha_s$- and $\beta$-caseins by $Ca^{2+}$ is the result of a reduction of the overall charge on the proteins; at neutral pH $\alpha_{s_1}$-casein has a charge of $-22$ units and $\beta$-casein a charge of $-12$ units. The $Ca^{2+}$ first binds the divalent phosphoserine residues and then, when these are fully occupied, the singly charged acidic residues. As the charge is reduced the repulsion between molecules is reduced and aggregation and precipitation begin to take place. An equilibrium is established between precipitated and non-precipitated material which is affected by $Ca^{2+}$ and casein concentrations. Even at high $Ca^{2+}$ concentrations, some non-precipitated material remains. The extent of precipitation of $\beta$-casein decreases rapidly with a reduction of temperature, and at 4°C no precipitate is formed. There are similar, but much less marked changes for $\alpha_s$-casein.

*κ-casein.* κ-casein is not precipitated by calcium ions, and stabilises aggregates of $\alpha_s$- and $\beta$-caseins against precipitation. In the presence of κ-casein, $\alpha_s$- and $\beta$-casein–$Ca^{2+}$ complexes form stable colloidal particles incorporating the κ-casein, the size of the particles depending on the proportion of κ-casein and the concentration of $Ca^{2+}$. *para*-κ-casein (residues 1–105) is a strongly hydrophobic peptide, whereas the glycomacropeptide (the remaining 64 amino residues) is hydrophilic, has an overall net negative charge and contains the sites of glycosylation. The stabilising effect of κ-casein may be achieved by the interaction of the hydrophobic *para*-κ-casein moiety with an hydrophobic 'core' of the $\alpha_s$- or $\beta$-casein–$Ca^{2+}$ complexes, and the interaction of the hydrophilic glycomacropeptide moiety with the surrounding solvent. The stabilising effect of κ-casein is lost if the two parts of the molecule are cleaved by chymosin: in the presence of other caseins the glycomacropeptide remains in solution but the *para*-κ-casein precipitates together with the other casein constituents.

Isolated κ-casein exists in solution in an equilibrium between monomers and soap-like micelles formed, presumably, by the interaction of hydrophobic portions of the micelles. The two cysteine residues of κ-casein are, however, located in the *para*-κ-casein fraction, and this offers the possibility of formation of intermolecular disulphide bonds with neighbouring molecules, and also with cysteine-containing serum proteins during heat treatments that lead to the denaturation of the serum proteins.

*Casein micelles.* The casein micelles of whole milk are particles with an average radius of about 100 nm, consisting of $\alpha_s$- and $\beta$-caseins stabilised by $\kappa$-casein in association with $Ca^{2+}$ and colloidal calcium phosphate, and may be regarded as extensions of the complexes described above. The micelles may contain many thousands of casein molecules but there is a size distribution and small micelles are most numerous. The stability of the micelles is partly due to the presence of the $\kappa$-casein, but also to that of colloidal calcium phosphate (see the section on minerals). Assembly of the micelles occurs mainly in the golgi vesicles of the secretory cell, but post-secretory association also takes place.

The internal structure of the micelles is not firmly established: many models have been proposed (Fig. 1), none of which is wholly satisfactory. Some models represent the casein micelle as an aggregation of 'submicelles' that are themselves limited aggregates of casein molecules. Alternatively, it is possible to define the micelle as the result of a polyfunctional polymerisation of $\alpha_s$-, $\beta$- and $\kappa$-caseins having $>2$, 2 and 1 functional moieties per molecule, respectively. Neither mechanism explains in detail the role of calcium phosphate. Under the electron microscope the micelles appear as approximately spherical structures with a graininess which may be due to a submicellar structure, or to inhomogeneous polymerisation.

Though in freshly secreted milk there is little soluble casein and nearly all of the casein is present as micelles, when the milk is stored at low temperature much of the $\beta$-casein dissociates from the micelles. Little of the $\alpha_s$- and $\kappa$-caseins passes into solution, and it has been assumed that these two components provide the structural framework of the micelles.

Casein micelles can be caused to aggregate by several factors, but proteolysis by enzyme action is the most commercially important cause. During cheese-making, the formation of curd is brought about by the action of the chymosin contained in rennet destroying the stabilising effect of the $\kappa$-casein. Chymosin is highly specific in its action, but other proteases with a more general action can hydrolyse $\kappa$-casein with the formation of *para*-$\kappa$-casein. Proteolytic enzymes of psychrotropic bacteria also split $\kappa$-casein and can cause gelation of ultra-high temperature treated milk during storage.

Heating of milk to 120–140 °C can also affect micellar stability, but the physico-chemical effects are not fully understood. Two reactions have been identified. First, at 80–100 °C, the serum proteins denature and bind to the casein micelles, principally to the $\kappa$-casein component (see section on serum proteins), and secondly, perhaps as a consequence of the first reaction, the micelles aggregate to form a gel or a precipitate depending on conditions.

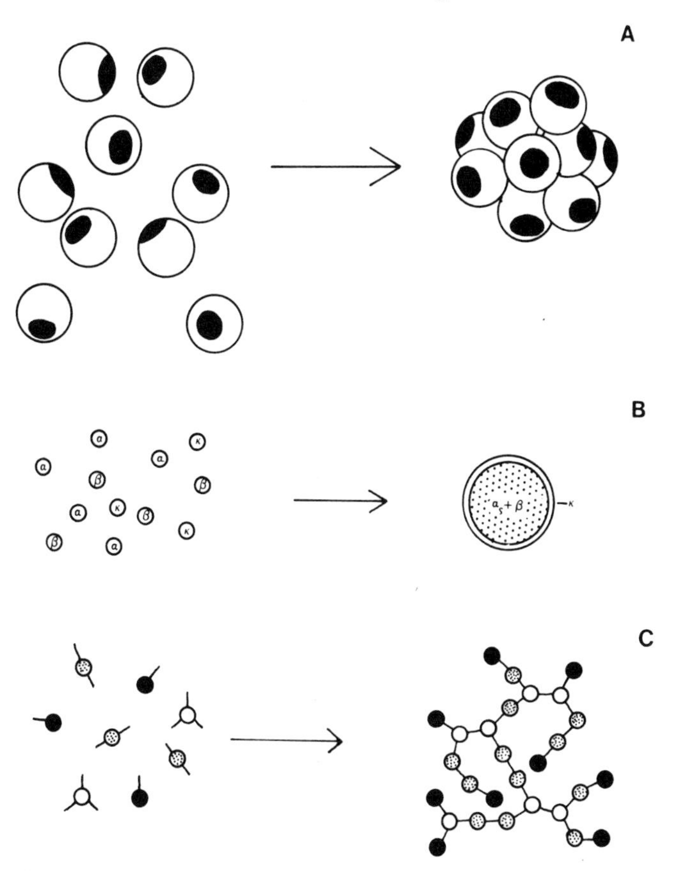


FIG. 1. Alternative concepts of casein micelle formation. A: aggregation of mainly hydrophobic subunits so that hydrophilic κ-casein regions (black) are on the outside of the aggregate. B: formation of a 'core' of hydrophobic $\alpha_s$- and β-caseins surrounded by a 'coat' of κ-casein. C: formation of a gel-like network of multifunctional units; monofunctional units are κ-casein, difunctional units are $\alpha_s$- and β-casein, and trifunctional units are $\alpha_s$-casein.

These reactions can occur during processing and assume importance in concentrated milk since precipitation then occurs at lower temperatures.

Another aspect of proteolysis, that does not affect micellar stability, is the production of γ-caseins and proteose-peptones from β-caseins by the proteolytic action of fibrinolysin (plasmin) which occurs naturally in small amounts in milk. The reaction is confined mainly to β-casein, possibly because of the easy dissociation of β-casein from the micelles, and is specific to certain positions in the molecule.

*Serum Proteins*

Relative to the caseins, the serum proteins have only a minor effect on the physical chemistry of untreated milk, but during heating their properties assume a greater importance, and the properties of the isolated whey proteins are of increasing commercial significance.

In fresh milk, the serum proteins are in solution and play no part in micellar structure. They have a more typically globular structure than the caseins and, in the native state, the many cysteinyl residues result in the formation of intra-molecular disulphide bonds which stabilise the protein structure. They lack phosphate groups, and do not bind $Ca^{2+}$ or aggregate strongly in the native state. At temperatures in excess of 80 °C the serum proteins denature, a process accompanied by extensive breaking and randomisation of the stabilising disulphide bonds. This process is especially marked in $\beta$-lactoglobulin, which possesses one free sulphydryl group that allows autocatalytic disulphide exchange reactions to be initiated.

At normal temperatures there is no evidence of interaction between $\beta$-lactoglobulin and $\kappa$-casein, although both possess free sulphydryl groups. At 100 °C, however, following denaturation of the $\beta$-lactoglobulin, interaction between $\beta$-lactoglobulin and $\kappa$-casein occurs, although the $\kappa$-casein remains fixed to the micelle. This alters the nature of the micelle surface, such that the stability of the micelles to heat and to enzyme action is affected. The heat-stabilities of forewarmed milks, that have been maintained at 90 °C for some time before raising the temperature to 120 or 140 °C, are significantly different from milks that have been heated rapidly, because of a more complete reaction between denatured serum proteins and $\kappa$-casein. In addition to the reaction with $\kappa$-casein, individual serum proteins will, when milk is heated, form gels by the same process of denaturation followed by intermolecular formation of disulphide bonds, leading, because of the number of cysteinyl residues in the proteins, to the formation of infinite networks of linked protein molecules.

Attempts have been made to incorporate the serum proteins into cheese by heating the milk, prior to renneting, to induce complex formation with $\kappa$-casein, but this has been found to inhibit aggregation, presumably because the bound serum protein stabilises the *para*-casein micelle. Serum proteins, especially when they have been denatured, can also act as emulsifying agents towards lipid, because of their ability to interact with hydrophobic particles on the one hand, and solvent molecules on the other.

*General*

The heating of milk also causes a general heat-induced proteolysis and, in addition, specific chemical modifications of the milk proteins. Three of these are of particular interest, the formation of lysino-alanine, the interaction with cyanate and the Maillard reaction with lactose (see section on lactose). Formation of lysino-alanine occurs through the $\beta$-elimination of phosphate from a phosphoseryl residue under alkaline conditions, followed by the reaction of the dehydro-alanine with the amino group of lysine.

```
        OH
        |
    O=P—OH
        |
        O               NH2                                    NH2
        |               |                                      |
       CH2            (CH2)4     ——→        CH2              (CH2)4
        |               |                   ||                 |
       CH               CH                  C                  CH
      /  \             /  \                / \                /  \
—HN      CO------HN       CO---       —HN    CO------HN          CO---

                    CH2——NH——(CH2)4
                     |          |
                    CH          CH
                   /  \        /  \
               —HN    CO------NH    CO---
```

Hydrolysis of the protein liberates the lysino-alanine.

Cyanate is present in solution in equilibrium with urea:

$$CO(NH_2)_2 \rightleftharpoons N^+H_4—CNO$$

and on heating reacts with lysyl amino groups to give homocitrulline.

```
                                                 HN    NH2
                                                   \\  /
                                                     C
       NH2                                           |
        |                                            NH
      (CH2)4                                         |
        |                                          (CH2)4
       CH       + N+H4—CNO  ——→                      |
      /  \                                           CH
  —HN     CO—                                       /  \
                                                —HN     CO—
```

**Minerals**

The interactions between $Ca^{2+}$, $Mg^{2+}$, phosphate, citrate and casein determine the ionic composition of milk, in addition to affecting the structure of the casein micelles. As previously described, the calcium ions bind to some of the caseins, but they also bind to phosphate and citrate, resulting in the formation of insoluble calcium phosphate and soluble calcium citrate complexes. The equilibria between the various ions and the properties of the complexes are such that the larger part of the citrate is present as Ca $Cit^-$, and approximately half of the inorganic phosphate exists as precipitated calcium phosphate, leaving only about 2 mM (out of some 30 mM) of the calcium in solution as the $Ca^{2+}$ ion. Similarly, much of the magnesium present is complexed by citrate, thus reducing the concentration of free $Mg^{2+}$ in solution.

Colloidal calcium phosphate appears to be an essential component of the natural micelle. Though it is possible to form micelle-like particles from casein and $Ca^{2+}$ alone, these are less stable than the natural micelle, and the inclusion of phosphate enhances stability. Theories as to the function of the calcium phosphate range from a role in the binding of the submicellar complexes, to that of acting as a buffer for calcium ions when the slow response to physical change of insoluble calcium phosphate would confer stability. No model so far proposed accounts fully, however, for the observed effects of colloidal calcium phosphate on micelle stability.

The soluble calcium concentration in milk correlates closely with soluble citrate concentration and there is evidence, for bovine milk, that secretion of $Ca^{2+}$ is linked with that of citrate. Models have been constructed, based on the separate interaction equilibria for the ionisation of phosphate and citrate and the binding of $Ca^{2+}$ to the different phosphate and citrate ions, that give results in good agreement with analyses of milks, not only from cows, but from other species also.

Phosphate, citrate and calcium salts are used widely as milk additives during processing. The anions complex additional $Ca^{2+}$, so reducing that available for binding to casein and thus stabilising the micelles against aggregation. Conversely, the addition of calcium reduces the stability of the micelle and may be used to enhance curd formation in cheese production.

**Lactose**

Milk is isotonic with blood plasma, and the colligative properties of milk (osmotic pressure, freezing point depression, boiling point elevation) show the same constancy as for other body fluids. Lactose makes a major contribution to these properties (accounting for approximately 50% of the osmotic pressure of milk) and variations in its concentration in milk are associated with inverse changes in the concentrations of other water-soluble constituents, sodium and chloride in particular.

Three solid forms of lactose exist, namely $\alpha$- and $\beta$-lactoses (anhydrides) and $\alpha$-lactose monohydride. In solution, an equilibrium mixture of the $\alpha$- and $\beta$-anomers is attained, the composition of which is dependent on the temperature; the ratio of $\beta$-lactose to $\alpha$-lactose is 1·65 at 0 °C and 1·58 at 25 °C. Lactose is one of the more insoluble of the common sugars; its solubility at 25 °C is only 17·8 g per 100 g solution, and this relatively low solubility can cause problems in processing. Thus, the concentration of lactose normally present in ice cream and in sweetened condensed milk exceeds the solubility limit, but the conditions of processing are so arranged that only very small crystals are produced, so small that the tongue cannot detect their presence.

Lactose is much less sweet than sucrose, and also less sweet than equimolar mixtures of its component monosaccharides, galactose and glucose. Like other reducing sugars, it can react with the free amino groups of proteins to give products that are brown in colour (the so-called Maillard reaction). The extent of the reaction depends on such variables as the concentration of lactose and protein, pH and the time and temperature of processing. The reaction is highly complex, but is thought to occur through the interaction of amino groups with aldehydes.

**Fat and Fat Globules**

Fat is present in milk in the form of globules of lipid surrounded by a phospholipid-rich layer, the milk fat globule membrane, which is derived during the secretion process from the apical membrane of the secretory cell. In freshly secreted milk, the globules are fairly large, with diameters of up to 6 $\mu$m but, as with casein micelles, there is a distribution of size. The surrounding membrane gives stability to the globules and, since the hydrophobic lipid is not in direct contact with the aqueous phase of the milk, the fat remains well dispersed throughout the milk. Skimming or separation of milk makes use of the density differences between the fat and the aqueous phase, and does not constitute a true phase separation.

In freshly secreted milk, there is sufficient membrane to cover most of the lipid surface. Mechanical stress (homogenisation in particular), that reduces globule size and increases surface area, makes the lipid more accessible to lipolytic action as the protective phospholipid coat is lost. Other physical changes also take place in the vicinity of the globule. The milk proteins (principally the caseins and, to a much lesser extent, the serum proteins) bind to the newly formed fat surface and act as emulsifying agents to maintain the fat in suspension as discrete particles. (The caseins are excellent emulsifiers because of the presence of strong hydrophobic and hydrophilic moieties.) The tendency of caseins to bind to the fat surface is so strong that not only can intact micelles act in this way, but the micelles may be broken up to give a layer of 'submicellar' casein particles. The replacement of the original fat globule membrane structure by a casein layer affects the behaviour of the fat globules as the casein layer retains many of the properties of the original material. However, the properties of the casein are altered in certain ways, and the gelation and precipitation of casein when milk is heated or renneted is reduced by homogenisation.

Coalescence of fat globules occurs when the globule membrane is disrupted in the absence of sufficient protein to maintain a stable emulsion, as in the case of separated cream and under such circumstances, the physical properties of the lipid become important. (When the lipid is contained within the original membrane, or is coated with casein, the effect of the lipid on the physical properties of other milk constituents is virtually independent of the nature of the lipid itself.) The most important physical property of the lipid is the melting point of the individual lipid molecules that make up the bulk phase. The melting point depends on the nature of the hydrocarbon side-chains of individual triglycerides, and is increased by increasing chain length and decreased by unsaturation of the side chains. The lipid is normally liquid at temperatures of 40 °C and above but will start to crystallise as the temperature is decreased so that, for average milk fat, half is solidified at 12–15 °C. This crystallisation of lipid, and possibly of the phospholipid of the membrane, may be a cause of the breakdown in fat globules that can occur on cooling.

The two most important chemical reactions in which the triglycerides of milk fat participate are hydrolysis (lipolysis) and autoxidation. The natural lipoprotein lipase of milk acts preferentially at the *sn*-3 position of the triglyceride which, in bovine milk, contains a high proportion of fatty acids of short chain-length (see Table I). These low molecular weight acids are volatile and give rise to unacceptable odours (particularly butyric acid); microbial lipases are less specific and give more general selection of fatty acids. Autoxidation occurs by the formation of a free radical at a methylene group adjacent to a double bond, and by subsequent reaction with oxygen.

$$-CH_2-CH{=}CH- + R^* \rightarrow RH + -C^*H-CH{=}CH- \qquad (1)$$

$$-C^*H-CH{=}CH- + O_2 \rightarrow -\underset{\substack{|\\ O-O^*}}{CH}-CH{=}CH- \qquad (2)$$

$$
\begin{array}{l}
\underset{\displaystyle \mathrm{O{-}O^*}}{\underset{|}{\mathrm{-CH}}}\mathrm{-CH{=}CH-} + \mathrm{-CH_2{-}CH{=}CH-} \longrightarrow \\
\qquad\qquad\qquad \underset{\displaystyle \mathrm{O{-}OH}}{\underset{|}{\mathrm{-CH}}}\mathrm{-CH{=}CH-} + \mathrm{-C^*H{-}CH{=}CH-} \qquad (3)
\end{array}
$$

Where * denotes a free radical.

Once started, the process is autocatalytic. The saturated nature of milk fat makes it fairly resistant to autoxidation, but nevertheless care should be taken to exclude catalysts for the reaction, such as copper. With milk fat containing a high proportion of polyunsaturated fatty acids, obtained by feeding cows vegetable oils protected against ruminal hydrogenation, it may be necessary to add an anti-oxidant at milking to prevent the formation of off-flavours associated with autoxidation.

## MILK PROCESSING

The aim of the modern milk processor is to produce, as economically as possible, a material that appeals to the consumer and has a reasonable shelf-life; it is impossible to define shelf-life in absolute terms since it varies with the nature of the product. It is the intention of this section, therefore, to summarise the various processes to which milk is subjected, and to indicate the reasoning that lies behind the choice of processing conditions. Further details of these processes can be found in subsequent chapters, and hence the following accounts should be regarded as no more than process outlines.

### Pasteurised Milk

Approximately 50 % of the milk produced in the UK is sold to the consumer as liquid milk, and virtually all of that liquid milk is pasteurised. The original object in pasteurising milk was to prevent it acting as a carrier of human pathogens, particularly those causing tuberculosis and brucellosis. The process was introduced at a time when milk was collected from farms in churns, at ambient temperatures, and the dominant bacterial flora was composed of lactic acid producing, Gram-positive bacteria, which caused souring. Over the last 20 years, the churn has been replaced by the refrigerated bulk tank, and milk is now collected from the farm and transported at low temperatures. The dominant bacterial flora has consequently changed to a psychrotrophic one, mainly Gram-negative rods, and spoilage now results not from souring, but from various off-flavours, e.g. metallic or tallowy. In addition to these changes in the storage and collection of milk, tuberculosis has been completely eliminated from the national herd, and brucellosis is very close to eradication.

Despite these changes, there are still significant risks in consuming raw milk (e.g. salmonellosis), and milk for the liquid market is still pasteurised. The process does ensure human health and, by greatly reducing bacterial numbers, extends the shelf-life of the milk. The actual heat treatment accorded the milk has also changed. Originally, the milk was held at a temperature of 64 °C for 30 min, but this method has been supplanted by the high temperature–short time (HTST) technique, in which the milk is held at 74 °C for not less than 15 s.

Spoilage of pasteurised milk is due to one of two causes: (1) post-pasteurisation contamination, or (2) the growth of organisms that have survived heat treatment. Generally, if the flora at spoilage is dominated by Gram-negative rods, post-pasteurisation contamination is indicated, whereas the presence of Gram-positive organisms suggests that spore-forming bacteria have survived heat treatment.

**Pasteurised Cream**

The different types of pasteurised cream, and their fat contents, on sale in the United Kingdom are shown in Table V. Additives are not permitted,

TABLE V

VARIOUS TYPES OF PASTEURISED CREAM, AND THEIR FAT CONTENTS, ON SALE IN THE UNITED KINGDOM

| | *Clotted cream* | *Double cream* | *Whipping cream* | *Whipped cream* | *Single cream* | *Half cream* |
|---|---|---|---|---|---|---|
| Minimum content of milk fat (%) | 55 | 48 | 35 | 35 | 18 | 12 |

except in the case of clotted cream where the addition of nisin is allowed, and whipped cream, which, in addition to sugar, may contain alginate (or a mixture of sodium bicarbonate, tetrasodium pyrophosphate and alginic acid), sodium carboxymethyl cellulose, carrageenan, gelatine and nitrous oxide.

The process shown in Fig. 2 for the production of double cream is typical of that used for the preparation of all creams other than clotted cream, in that, following separation, the cream is standardised to the desired fat

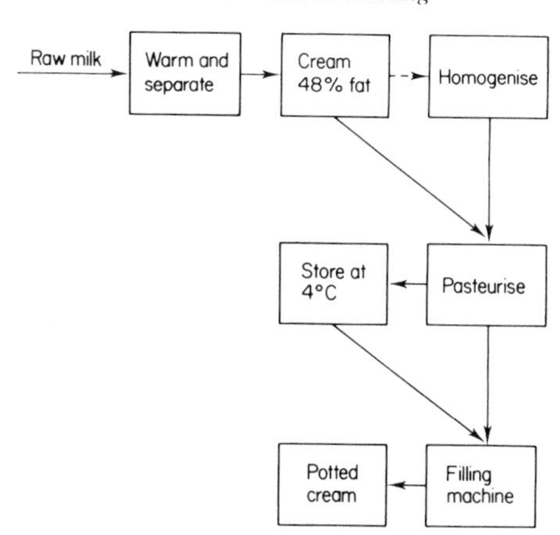


FIG. 2. The production of double cream.

content and then pasteurised. This latter process is now carried out using an HTST treatment, and a typical heat treatment would be 74 °C for 15 s, or 80 °C without holding; other intermediate temperature/time combinations may be employed.

Whether or not the cream is homogenised depends on the nature of the product. Single cream and half cream are almost invariably homogenised in order to prevent serum and fat separation, and to give an acceptably high viscosity. Double cream may or may not be homogenised, depending on the desired viscosity. If a very thick cream is required, it can be produced by a relatively low pressure homogenisation; alternatively, a similar effect may be achieved by cooling the cream to 27 °C as it leaves the pasteuriser, and then allowing it to cool very slowly to 5 °C or less. Slow cooling, with the attendant risk of bacterial growth, necessitates that the utmost attention be given to the hygienic condition of the plant. Homogenisation is not generally applied to whipping cream, unless the product has been subjected to an ultra-high temperature (UHT) procedure.

Some manufacturers favour homogenisation before pasteurisation, whereas others prefer it after pasteurisation. One aspect that may mitigate against prior homogenisation is the presence in milk of the natural lipoprotein lipase. This enzyme is capable of considerable activity in milk, but is prevented from reaching its substrate, the triglyceride, by the milk fat globule membrane. Homogenisation, however, increases the surface area of the milk fat so greatly that it cannot be covered by the membranous material and, although milk proteins are absorbed onto the residual surfaces, they are not so protective against the lipolytic activity. As lipoprotein lipase activity is destroyed by pasteurisation, a lower level of free fatty acids is to be expected in products in which homogenisation occurs after pasteurisation.

The distinguishing characteristic in the manufacturing process of clotted cream is that it is heated in shallow, jacketed tanks, either alone or floating on a layer of skim-milk or whole milk. The cream is heated to 77–85 °C for 45–70 min, during which time a crust forms. It is then cooled and, after approximately 12 h at 7 °C, the cream may be packed.

The conditions of manufacture of clotted cream are conducive to microbial growth, in that the product is slowly cooled and then stored in open trays. Extreme care must be taken to ensure that the utensils and equipment are sterilised before use, and ideally processing and storage should be carried out in slightly pressurised rooms, the air supply having passed through a bacteriological filter. The permitted addition of nisin, composed of a series of closely related antimicrobial polypeptides produced by strains of *Streptococcus lactis*, interferes with the germination of spores, and hence is effective against spoilage by Gram-positive organisms.

**Butter**

The present UK legal requirement for butter is that it should contain not more than 16% water and not less than 80% milk fat; the term 'butter' is applicable only to a product in which all the fat is derived from milk.

In cream, water is the continuous phase and fat the disperse phase, whereas in butter the roles are reversed. Traditionally, this phase inversion was carried out by the processes of churning and working. In the initial phase of churning, partially solidified milk fat globules are brought together forming small grains, and further churning causes an increased flow of liquid fat from the globules leading to the formation of larger grains. Eventually the butter separates ('breaks') from the buttermilk, which can be drained away allowing the butter to be washed. The homogeneity of the product is then ensured by a kneading process called 'working'. During working, adjustments can be made to the moisture level, and salt may be added. The product has a continuous fat phase, consisting of fat globules that have been forced together, fragments of globules and fat squeezed from the globule. In the ideal emulsion, the aqueous phase would consist of small discrete droplets, and whilst this is true of the major portion of the water in butter, a small amount exists in the form of minute continuous channels. The susceptibility of the butter to microbial spoilage appears to be closely related to the proportion of water that is present in these continuous channels.

The modern conventional churn encompasses the processes of churning and working in the same vessel. Churning is accompanied by the normal rotating motion, and working is effected by the butter falling against steel bars fixed within the churn. However, even the modern churn has, in commercial practice, been replaced by the continuous buttermaker. In its most rudimentary form, the continuous buttermaker consists of a cylinder within which there is a high speed dasher. The incoming cream is thrown against the blades, and churning to butter occurs within 2 s. The butter grains are pushed out of the cylinder by means of a screw, and the buttermilk run off. Working is performed by the action of the screw as the butter is extruded from the machine. Manufacturers are now tending to put butter directly into retail packs as it comes from the continuous maker, rather than storing it in the traditional 25 or 50 kg blocks. In the latter system, the butter was reworked immediately before retail packaging, a process which led to a less hard product; direct packaging, by avoiding this softening stage, can yield a butter that is less spreadable than its traditional counterpart.

Improving the low-temperature spreadability of butter is a subject that has received much attention since the introduction of the luxury tub-type margarines. A great deal of interest has been shown, particularly on the continent, in various time–temperature combinations of treating the cream prior to churning, i.e. variations of the so-called Alnarp process. This type of processing does lead to a more spreadable butter, but the effect is achieved by setting up a metastable crystalline state. If the butter is held constantly at refrigeration temperature, the metastable state persists but, on warming, the crystalline content of butter gradually decreases, and subsequent cooling produces the stable crystalline form. As a consequence of temperature cycling, therefore, butter prepared by the Alnarp process loses its enhanced spreadability.

The main determinant in the spreadability of butter is its fatty acid composition, and although the fatty acids vary markedly with season, their essentially saturated nature is maintained. Butter prepared from milk when the cows first go out to grass usually has the best low-temperature spreadability, but is nevertheless much harder than the tub margarines; winter milk can yield a butter that cannot be spread at refrigeration temperature. The diet of the cow can be manipulated to increase the unsaturated fatty acid content of the milk fat, but it appears unlikely that methods of producing more spreadable butter that are dependent on farming practice will be commercially viable. Given the dominant role of fatty acid composition in determining spreadability, any future attempts to improve the latter parameter will more likely involve the addition of vegetable oils at the manufacturing stage. The resultant product cannot, of course, satisfy the present legal definition of butter. An alternative, but more expensive, solution is to fractionate the milk fat to be used in butter manufacture according to its melting point, and then to combine those fractions having the desired properties.

Some manufacturers store frozen cream from springtime and use it in admixture throughout the year. This approach removes the very large seasonal variation in spreadability, but the economics require careful consideration.

To ensure a product of high hygienic quality, the cream is pasteurised prior to churning. Whilst normal HTST pasteurisation is quite adequate for reducing bacterial numbers, some advantages are claimed for using a 'vacreator' system, which operates at a reduced pressure. In particular, vacreation removes feed taints and butyric acid (which results from the action of lipoprotein lipase) from the milk fat. After pasteurisation, the cream is cooled and held for some time to allow crystallisation of the milk fat to occur. If a lactic butter is required, the necessary organisms are introduced when the pasteurised cream is partially cooled; when acid development has reached the required stage the cream is further cooled to allow the crystallisation process to proceed. When the cream is suitably aged it enters the continuous buttermaker. The retail packs are very soft, but firmness increases with time due to the process of setting. The greater

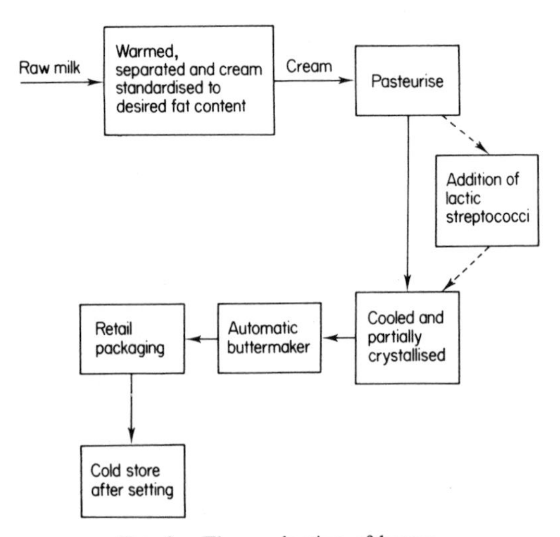


FIG. 3. The production of butter.

part of the setting occurs within 24 h, but it can be delayed if the butter is immediately transferred to a cold store. The manufacturing process for butter is shown in summary form in Fig. 3.

Of all milk products, butter is the least conducive to bacterial growth. This is especially true of salted butter, in which the aqueous phase is a brine solution, and, to a lesser extent, of lactic butter. Surface taints are sometimes observed, which originate from bacterial contamination of the wash water. However, even though most butter is made in the summer months, it can be stored at deep-freeze temperatures (*c.* −25 °C) for many months without any material change in flavour.

**Cheese**

A typical production process for Cheddar cheese, which accounts for 70 % of the cheese produced in the UK, is shown in Fig. 4.

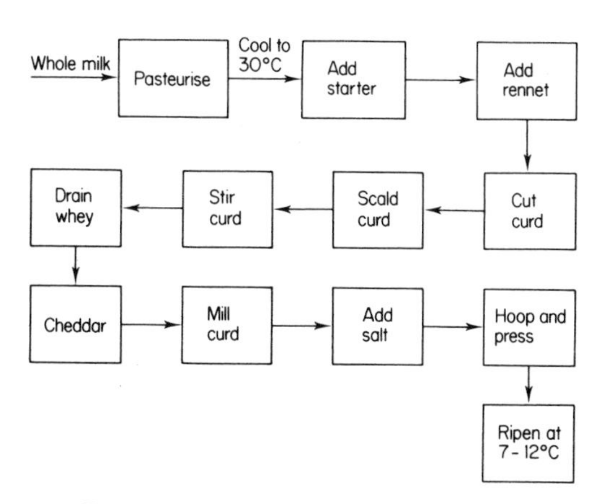


FIG. 4. The production of Cheddar cheese.

The raw milk undergoes a fairly mild HTST pasteurisation, typically 70 °C for 16 s, and is then cooled to 30 °C. A starter culture, consisting of one or more strains of lactic acid producing bacteria, is then introduced at a concentration of *c.* 2 %, by volume. When the acidity has reached a suitable level, rennet (at about 0·03 %) is introduced. Curd development proceeds for about 40 min, at which point the curd is cut into small pieces. These pieces are kept afloat in the whey by gentle stirring whilst the temperature is slowly raised to 39 °C—the process of scalding. This process allows the development of the desired acidity and texture in the curd, which is then separated from the whey. The particles are then allowed to coalesce under their own weight, an action which, with the further expression of whey, comprises the cheddaring process. After cheddaring, the curd is milled, i.e. cut into small rectangular pieces, and salt added (2–3 kg per 100 kg curd). When dissolution of the salt is complete, the curd is packed into hoops lined with cheesecloth and pressed for periods of up to 24 h. In small sizes (0·5 kg), the cheese is dried for a few days at 10–15 °C to form a rind, and then dipped in cheese wax before curing. In larger sizes, rindless cheese is made by sealing the material from the presses in plastic film. The ultimate stage in the cheesemaking process is curing, i.e. storage at 7–12 °C for periods between 3 months and 1 year.

It is during storage that the compounds associated with the typical flavour and aroma of Cheddar cheese are formed, and volatile sulphur compounds, such as hydrogen sulphide and methanethiol, contribute to both properties. They result from the formation of reduced sulphur compounds which decompose to hydrogen sulphide; a portion of the gas reacts with the methionine residues to produce methanethiol. The development of these compounds can be accelerated by the addition of such chemicals as cysteine, glutathione and dithiothreitol.

Non-volatile flavour components are produced by the breakdown of milk proteins. Attempts are currently being made to accelerate the production of these compounds by the addition of proteases and peptidases to the curd, for the possibility of producing a cheese of mature flavour with reduced ripening times is economically attractive.

The cheesemaker has a two-fold aim, namely to produce a cheese of high quality, whilst obtaining a high yield from a given quantity of milk. The compositional factors that dominate cheese yield are the proportions of casein and fat. Casein constitutes the matrix of the cheese, whereas the fat may be regarded as an inert filler, and there is an optimum casein:fat ratio in milk for Cheddar cheese making, the accepted value being 0·7. As the ratio falls, there is a greater probability of fat losses in the whey, whereas as it rises, the water content of the cheese increases. Although the legally permitted maximum water content of Cheddar cheese is 39 %, quality considerations dictate that the consumer receives a product of somewhat lower water content—generally, the aim is to produce a Cheddar cheese containing approximately 35 % water.

In addition to the major effects of casein and fat, other milk constituents play a minor role in determining yield. For example, the amount of calcium present influences the physical structure of the protein matrix and affects fat losses in the whey; hence the proposal that milk for cheese manufacture should be standardised to the optimum casein:fat ratio might not result in a fixed weight of cheese per unit volume of milk.

Hygienic factors may also have a slight effect, for it has been shown that, at very high bacterial loads, proteolytic exo-enzymes can cause sufficient damage to the casein to reduce cheese yields. However, the commercial milk supply rarely reaches the required level of contamination. The concentration of somatic cells typically found in cheese milk is associated with slight increases in the proportion of soluble casein, which again can adversely affect cheese yields.

The more important effect of hygienic factors is to be found in relation to the off-flavours that develop in cheese during storage. The ability of the exo-enzymes produced by psychrotrophs to retain their activity after heat treatment can lead to unwanted degradation during storage. For example, milk with a psychrotrophic count exceeding $10^6$ organisms $ml^{-1}$ may produce sufficient lipase to cause rancidity of the cheese during storage.

## Full Cream Evaporated Milk

The UK specification for full cream evaporated milk is that it should contain not less than 9% milk fat and not less than 31% total solids. Permitted additives, which confer stability on the milk protein so that it may withstand in-can sterilisation, include $NaH_2PO_4$, $Na_2H_2PO_4$ and $Na_3PO_4$, $Na_3Cit$ and $H_3Cit$, $NaHCO_3$ and $KHCO_3$ and $CaCl_2$. The various stages in processing are shown in Fig. 5.

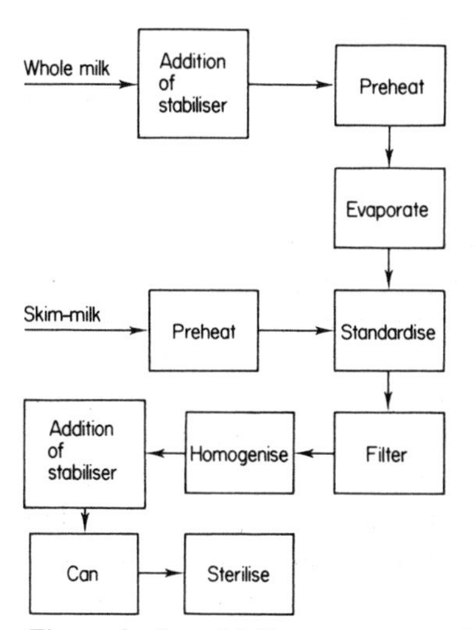


FIG. 5. The production of full cream evaporated milk.

Stabiliser is added to the raw milk, which is then subjected to heating. This pre-heating stage is intended to denature the serum proteins, which increases the stability of the product during sterilisation. Conventional evaporation then follows to produce a milk that has a higher total solids than is required in evaporated milk, and skim-milk concentrate (also preheated) is added to reduce the total solids and fat contents to the desired levels. The product is then filtered and homogenised; an inferior evaporated milk is obtained if the homogenisation is carried out prior to concentration. Samples are then tested for their ability to withstand the sterilisation process. If necessary, more stabiliser can be added at this stage, as long as the total addition does not exceed the legally permitted amount. The product is then sealed in cans, which are sterilised, e.g. at 120 °C for 10 min.

Because of the severity of the heat treatment and the consequent Maillard reaction, evaporated milk is darker in colour than is pasteurised milk. Despite the benefits gained from denaturing the serum proteins and the addition of stabilisers, difficulties during the sterilisation process can be encountered periodically if the milk supply is pronouncedly seasonal. Even when this lactational effect is not very marked, stability problems can arise during the storage life of the evaporated milk (especially in the tropics), which may be related to the season of milk production. Gelation of canned milk has occasionally been observed to be due to contamination by *Bacillus subtilis*, and cases of heat-resistant organisms, such as *Bacillus cereus*, surviving sterilisation to produce off-flavours (and odours) have been reported. Generally, however, the bacterial problem is virtually non-existent, and most problems encountered during storage are due to chemical or physical changes in the milk components rather than to microbial action.

**Skim-Milk Powder**

Virtually all the skim-milk powder currently produced in the UK is obtained by spray drying. The ways in which different types of powder may be produced are shown in Fig. 6.

A number of products of quite different properties can be obtained from the same starting material, skim-milk. Either the integrity of the whey proteins can be preserved (a low-heat powder), or they can be completely denatured (a high-heat powder). The process of agglomeration can be superimposed on both low- and high-heat powders to produce readily soluble products, i.e. instantised powders. The powder is made with the ultimate end-use in mind. Thus, for incorporation in a bread dough, a high-heat powder is necessary; incorporation of a low-heat powder leads to an

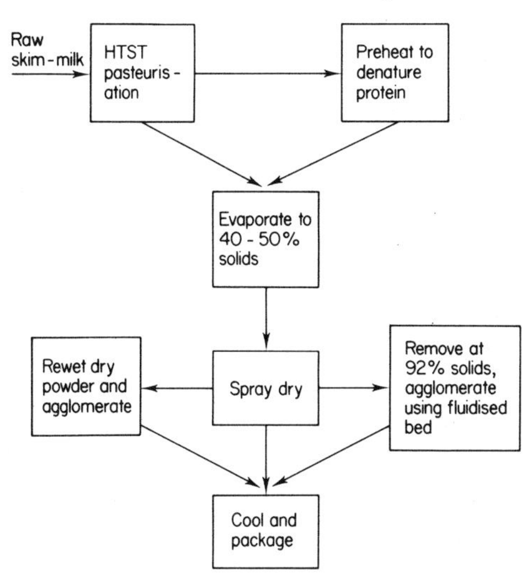


FIG. 6. The production of skim-milk powder.

unacceptable loss of loaf volume and a poorer crumb structure. On the other hand, powder destined for use as a coffee whitener should preferably be a low-heat powder, in order to minimise insoluble residues in the coffee cup.

The first treatment that the skim-milk receives is a normal HTST pasteurisation, e.g. 72 °C for 15 s. If a low-heat powder is required, the pasteurised skim-milk goes directly to the evaporators, but if a high-heat powder is required, it goes to a tubular heat exchanger at a temperature of 85 °C for a holding time of 20 min (these are representative conditions), and then to the evaporators.

The skim-milk is taken from the evaporator at a total solids content of 45–50 % and introduced into the spray drier. It is cheaper to remove water in the evaporator than in the drier, and hence a high total solids content is desired at this stage. As the total solids content increases, so the viscosity increases, and to prevent pumping difficulties the temperature must be raised. However, at higher temperatures, the probability of protein interactions, that could lead to deleterious changes in product quality, e.g. a decrease in solubility, increases, and hence the upper concentration limit on the material coming from the evaporator is *c.* 50 %.

When the concentrate reaches the drier it is atomised. Heating can be either direct, i.e. gas is burned in the incoming air, or indirect, i.e. the heat comes from electrical heaters or from steam-based heat exchangers. Milk powders made using direct heating may have higher levels of nitrates and nitrites, a point of particular importance in the case of infant formulations. Steam shielding of the gas flame lowers the nitrate and nitrite levels, but introducing water to the drier chamber somewhat reduces the gain in thermal efficiency associated with direct firing. The design of driers is such that contact between the milk spray and the hot air is maximised, allowing for a very efficient removal of moisture from the spray droplets. Although the air inlet temperature may be up to 180 °C, the temperature of the spray particle does not exceed the dew point, and hence there is no danger of damage to the protein. Only when some 96 % of the water has been removed does the temperature of the particle increase beyond the dew point, but by that time, it has passed from the hot zone. The bulk of the powder can be removed directly from the main chamber, but the fine particles are extracted from the exhaust air by a cyclone and in some cases the air is finally scrubbed. The resultant powder can be cooled and packed.

The drier is normally designed to minimise contact between particles when they are sufficiently wet to stick to one another. However, for an agglomerated (instant) powder, conditions in the drying chamber are arranged to facilitate this sticking together. The product is then taken from the drier at approximately 8 % moisture, and the final drying performed using a fluidised bed. This process gives the so-called 'straight through' instantised powder. In the alternative procedure, powder at 4 % moisture is taken from the drier and subsequently treated with steam (the process of re-wet agglomeration), before again being dried on a fluidised bed.

Currently, much interest is being shown in dairying countries in the production of powder for recombination. This concept envisages that skim-milk powder could be exported to Third World countries, reconstituted, a locally produced fat added, and the resultant homogenised mixture sold either as an in-can sterilised product (equivalent to full cream evaporated milk), or as a UHT concentrate. The properties of the powder necessary to yield high quality recombined products are presently being defined.

Bacteriological problems are generally associated only with low-heat powders, for the concentration of Gram-positive spores surviving heat treatment can adversely affect the products into which the powder is to be put.

### Ultra-High Temperature (UHT)

Ultra-high temperature treatment is designed to confer a shelf-life of several months. In the production of UHT skim-milk, two distinct processes, shown diagramatically in Fig. 7, are involved: first, the heat treatment to destroy contaminating bacteria, and secondly, aseptic packaging of the sterilised milk. The two processes may be made independent by the introduction of an aseptic balance tank which holds the sterilised milk prior to packaging. In the case of fat-containing products, homogenisation, again carried out under aseptic conditions, generally follows the UHT process. Technically, it is simpler to carry out the homogenisation prior to UHT, but in some designs of plant, the heat treatment causes clumping of the homogenised fat globules, thus negating the advantages of homogenisation.

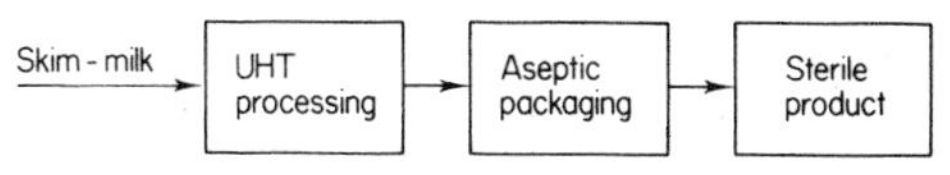


FIG. 7. The production of UHT skim-milk.

The objective of UHT is to produce a material that, in commercial terms, may be regarded as bacteriologically sterile but which retains the desirable characteristics of the fresh counterpart, i.e. nutritive value, colour and organoleptic properties. These, to some extent conflicting aims can be achieved in the case of milk products by the application of a temperature of not less than 125 °C for not less than 1 s, according to the definition of UHT processing recommended by the International Dairy Federation. In practice, the temperature range used is 135–150 °C, the holding time being several seconds.

The activation energy for the thermal death of the spores of *Bacillus stearothermophilus*, which are among the most heat-resistant cells likely to be found in milk, is almost three times greater than that for the Maillard reaction, so that an increase in temperature produces an increase in the ratio of sterilisation/browning. Also, with increasing temperature of heat treatment, the time required to induce a given level of spore denaturation decreases, e.g. an increase of only 1·5 °C in the UHT range corresponds approximately to a halving of the holding time. The difference in activation energies in the two processes, allied with the short holding times associated with high temperatures, yields products that compare very well in colour with their pasteurised counterparts.

Basically, there are two forms of UHT processing, which depend respectively on (1) indirect or (2) direct heating of the raw material. In the case of indirect heating, heat exchangers are used, which may be either plate or tubular in design. This equipment is thus similar to that used in normal HTST pasteurisation. In the indirect system, high pressure steam is mixed with the milk, either by injecting steam into the milk, or by infusing milk into the steam. Both of these operations add water to the product, but at a subsequent stage, the product is brought to the desired total solids level by an evaporative cooling process.

The direct heating system is more expensive, both in the capital investment involved and in running costs. It does, however, have the advantage of a very high heat transfer rate, so that the desired processing temperature is reached virtually instantaneously. (In both direct and indirect UHT systems, the milk is heated to *c.* 85 °C and held at this temperature for 40–60 s before undergoing the UHT stage.) In the case of indirect heating, some 5–7 s are required to reach the desired temperature. During this heating process, the milk becomes somewhat unstable, leading to the formation of deposits on the heat exchanger. The consequences of this deposition are (1) poor heat transfer, which may make attainment of the desired operating temperature difficult, and (2) constriction of the flow passages, leading to undesirably high pressures within the system.

Whilst both direct and indirect heating systems give products with similar shelf-lives (and are, therefore, equally effective in their bacteriological action), it is claimed by some that the organoleptic properties of the products from direct UHT processing are superior. However, the entire area of flavour is so complex that it is difficult to make a quantitative judgement on the merits of the two heating systems. For example, denaturation of $\beta$-lactoglobulin occurs during the prewarming stage, exposing sulphydryl groups, which then give the product a typical sulphurous smell and flavour. Milk heated by the indirect system contains rather more oxygen, unless a de-aerator has been introduced into the process line, than does milk undergoing direct heating, and hence the sulphurous characteristics are more rapidly removed by oxidation. Because of this oxidation reaction, the acceptability of UHT milk tends to increase for the first few weeks, but the disappearance of the flavour characteristics associated with the sulphydryl groups unmasks the true UHT flavour, which becomes more pronounced with increasing time of storage. The time of storage necessary to attain maximum consumer acceptability varies with temperature, but a period is always required to remove the sulphurous characteristics.

It has recently been suggested that the initial storage period could be avoided if the sulphur-containing compounds, giving rise to the cooked flavour, were modified after UHT processing. The enzyme, oxidase, which catalyses the reaction:

$$2RSH + O_2 \rightarrow RSSR + H_2O_2$$

has been suggested for this purpose, and by passing milk from UHT processing over a solid support on which sulphydryl oxidase was immobilised, the cooked flavour was entirely removed from the freshly packaged milk. Indeed, it has been claimed that UHT milk produced in this manner is indistinguishable in taste from pasteurised milk.

This type of enzyme treatment removes the need for storing UHT milk prior to sale. A second type of enzymic modification, which takes advantage of the storage period, results in a sweet milk. This is achieved by introducing aseptically, during packing, a small amount of $\beta$-galactosidase, which hydrolyses the lactose to glucose and galactose.

Whilst the presence of oxygen apparently helps the organoleptic acceptability, it is detrimental to the nutritional value of the product. Negligible loss of vitamins occurs during the UHT process, but the presence of oxygen during storage leads to the gradual oxidation of ascorbic and folic acids. The removal of added water in the direct heating process can reduce oxygen levels to values at which the oxidative loss of vitamins does not occur; the beneficial effect is often nullified, however, by packing the UHT product in a container permeable to oxygen. Very few processors use de-aeration with the indirect UHT system, and hence the loss of vitamins is inevitable, irrespective of the nature of the packaging material.

In addition to the changes in organoleptic and nutritional properties that occur during storage, a change in the consistency of UHT milk is observed, leading eventually to gelation. It is now thought that this type of deterioration is due to residual enzyme activity. Exo-enzymes, particularly proteases and lipases, produced by psychrotrophic bacteria commonly found in milk are extremely resistant to thermal denaturation. In some cases, more than 50 % of the enzyme activity can survive UHT processing, despite the organisms themselves being destroyed. Thus, although UHT processing can give a sterile product from raw milk containing a substantial bacterial load, shelf-life can be related to the hygienic quality of the starting material. It has, for example, been shown that milk containing a strain of *Pseudomonas fluorescens* at a level of approximately $5 \times 10^7$ cfu $ml^{-1}$† gelled 10–14 days after UHT processing; at a count of $3 \times 10^6$ cfu $ml^{-1}$, gelation did not occur until 56 days after UHT processing.

Currently the sale of UHT milk products in the UK is relatively small. The British consumer is, at the moment, satisfied with the daily delivery system of pasteurised milk, but as transport and energy costs rise, the advantages of a long-life product that does not require refrigeration become more apparent. The knowledge already exists to produce a UHT milk that is nutritionally equivalent to its pasteurised counterpart, but problems still remain in terms of organoleptic acceptability.

† Colony forming units per millilitre.

## REFERENCES AND BIBLIOGRAPHY

CHRISTIE, W. W. (1978) *Progr. Lipid Research*, **17**, 111.
IDF (1979) The physics and chemistry of milk proteins, *J. Dairy Research*, **46**, 161.
JENNESS, R. (1974) In: *Lactation vol. III*, Larson, B. L. and Smith, V. R. (Eds.), Academic Press, New York.
LAMPERT, L. M. (1975) *Modern Dairy Products*, 3rd edn., Food Trade Press, London.
MULDER, H. and WALSTRA, P. (1974) *The milk fat globule: emulsion science as applied to milk products and comparable foods*, Commonwealth Agricultural Bureaux, Centre for Agricultural Publishing and Documentation, Wageningen, The Netherlands.
PATTON, S. and JENSEN, R. G. (1975) *Progr. in the Chem. of Fats and Other Lipids*, **14**, 163.
WEBB, B. H., JOHNSON, A. H. and ALFORD, J. A. (1974) *Fundamentals of Dairy Chemistry*, The AVI Publishing Co. Inc., Westport, Connecticut.
WHITNEY, R. MCL., BRUNNER, J. R., EBNER, K. E., FARRELL, H. M., JOSEPHSON, R. V., MORR, C. V. and SWAISGOOD, H. E. (1970) *J. Dairy Sci.*, **59**, 795.

# 2

# Micro-Organisms Associated with Milk

A. GILMOUR and M. T. ROWE
*Department of Agricultural and Food Bacteriology,*
*Queen's University, Belfast, UK*

Milk, in addition to being a nutritious medium, presents a favourable physical environment for the multiplication of micro-organisms and being an animal product subject to widely differing production methods, results in its contamination by a broad spectrum of microbial types.

In this chapter, therefore, a selective approach has been adopted with regard to the choice of species discussed in detail, although every attempt has been made to include genera usually regarded as important.

Rather than using certain common characteristics to organise the bacteria into groups, e.g. psychrotrophic, thermoduric, etc., a more taxonomic approach has been adopted and the bacteria have been divided into families according to *Bergey's Manual of Determinative Bacteriology* (*8th edn.*) (Buchanan and Gibbons, 1974). It should be noted that specific reference has not always been made when information has been taken from this source.

It must also be emphasised that the aim of this chapter is essentially to provide a reference dossier on those organisms associated with milk and milk products, and the presentation of the relevant information is, therefore, entirely functional in approach. If this slight limitation is accepted, then the pages that follow should enable anyone not familiar with the more common genera to become rapidly acquainted with their important characteristics.

## FAMILY PSEUDOMONADACEAE

This Family comprises four genera, viz. *Pseudomonas*, *Xanthomonas*, *Zooglea* and *Gluconobacter*. Only the genus *Pseudomonas* together with the genus *Brucella* (a genus of uncertain affiliation) are considered important in the dairy environment.

### General Characteristics of All Genera

Straight or curved rods (0·5–1 μm × 1·5–4 μm). Gram-negative. Motile by polar flagella. Catalase-positive. Usually oxidase-positive. Chemo-organotrophs. Metabolism respiratory never fermentative. Growth requirements in most cases extremely simple. Strict aerobes except for those species that can use denitrification as a means of anaerobiosis.

### The Genus *Pseudomonas*

The following species of this genus are worthy of further consideration.

*(i) Pseudomonas aeruginosa*

Can be found in soil, water and urinary tract infections and has been shown to cause mastitis (Malmo *et al.*, 1972). Diffusible fluorescent pigments, and a soluble phenazine pigment, pyocyanin, produced by most strains in suitable media (King *et al.*, 1954).

RNA homology group I (Pecknold and Grogan, 1973). Arginine dihydrolase present. Can be distinguished by the production of blue, red or brown pigment on King's A medium, ability to reduce nitrate and nitrite, growth at 42 °C but not at 4 °C and ability to utilise ethanol and mannitol as a carbon source in ammonium salt sugars media (King and Phillips, 1978). Sodium-2-ketogluconate formed from sodium gluconate in Paton's medium (Hendrie and Shewan, 1966). Some strains produce the bacteriocin, aeruginocin. Bacteriocins are proteins having a lethal effect on bacteria, and are produced by other bacteria usually closely related to the sensitive strain.

Nutritionally versatile. Obligately aerobic except in media with nitrate. Optimum growth temperature is approximately 37 °C.

*(ii) Pseudomonas fluorescens*

Found predominantly in water and soil, and commonly associated with spoilage of foods, particularly if refrigerated prior to consumption. Spoilage in milk and milk products due to psychrotrophy and elaboration of heat-stable extracellular hydrolytic enzymes. Cultures produce diffusible fluorescent pigments particularly on King's B medium. Some strains produce a non-diffusible blue pigment. No pigment is produced on King's A medium.

RNA homology group I (Pecknold and Grogan, 1973). Arginine dihydrolase present. Usually unable to reduce nitrate and nitrite. Growth occurs at 4 °C but not at 42 °C. Can usually utilise mannitol but not ethanol in ammonium salt sugars media (King and Phillips, 1978). Has been divided into four biotypes, chiefly according to pigment production, ability to denitrify, synthesis of levan from sucrose and ability to utilise a number of carbohydrates as carbon sources. Hamon (1956) and Hamon *et al.* (1961) have described bacteriocins (fluorocins) from this species, although Patterson (1965) was unable to demonstrate bacteriocinogeny for *Ps. fluorescens* or *Ps. ovalis*.

Nutritionally versatile. Obligately aerobic except for a few strains able to use nitrate as terminal electron acceptor. Optimum growth temperature is approximately 25–30 °C.

(*iii*) *Pseudomonas putida*

Isolated from soil and, in lower numbers, from water. Produces a diffusible fluorescent pigment particularly on King's B medium. No pigment is produced on King's A medium.

RNA homology group I (Pecknold and Grogan, 1973). Arginine dihydrolase is present. Unable to reduce nitrate and nitrite (King and Philips, 1978) though Buchanan and Gibbons (1974) state that nitrite may be produced from nitrate. Growth occurs at 4 °C but not at 42 °C (King and Phillips, 1978). Able to utilise ethanol but not mannitol in ammonium salt sugars media (King and Phillips, 1978). Strains of this species may produce a bacteriocin (Clarke and Richmond, 1975).

Nutritionally versatile. Obligately aerobic. Optimum growth temperature is approximately 25–30 °C.

(*iv*) *Pseudomonas maltophilia*

Most frequently found in clinical specimens, but also isolated from water, milk, and frozen foods. No fluorescent pigments are produced (Juffs, 1973). Colonies may be yellowish but colour is not due to carotenoid pigments.

RNA homology group V (Pecknold and Grogan, 1973). Arginine dihydrolase is absent. Reduction of nitrate to nitrite may occasionally occur (King and Phillips, 1978). Oxidase-negative or delayed oxidase-positive (after 10 s). No growth occurs at 4 °C but may at 41 °C. Able to hydrolyse aesculin, and possesses lysine decarboxylase. Generally able to utilise lactose (King and Phillips, 1978).

Not as nutritionally versatile as other species mentioned. Obligately aerobic except for those strains able to utilise nitrate as terminal electron acceptor. Optimum growth temperature is approximately 35 °C.

(*v*) *Pseudomonas cepacia*

Widely distributed in soil, but also found in urinary tract infections. No fluorescent pigments are produced and usually there is no colony pigmentation.

RNA homology group II. Arginine dihydrolase is absent. Nitrite may be produced from nitrate. Most strains unable to grow at 4 °C but grow at 41 °C. Able to use levulinate, mesotartrate and tryptamine as carbon sources for growth. Possesses lysine decarboxylase (King and Phillips, 1978).

Nutritionally versatile. Obligately aerobic except those strains able to use nitrate as terminal electron acceptor. Optimum growth temperature is approximately 30–35 °C.

(*vi*) *Pseudomonas fragi*

Commonly found in soil and water (Morrison and Hammer, 1941). Considered by Buchanan and Gibbons (1974) to have been incompletely described, but generally considered to be a member of the genus *Pseudomonas*. No fluorescent pigments are produced, and non-pigmented colonies are observed (Hendrie and Shewan, 1966). Some strains may produce a brown diffusible pigment (Lysenko, 1961). Arginine dihydrolase is present. Nitrate is not reduced. Able to grow at 5 °C but not at 42 °C (Lysenko, 1961). Able to utilise ethanol and maltose in ammonium salt sugars media (King and Phillips, 1978).

(*vii*) *Pseudomonas putrefaciens*

Found in water and soil. Spoilage agent in fish, milk and milk products, e.g. causes surface taint of butter (Derby and Hammer, 1931). This organism has been placed in this genus for convenience though it does possess some properties which disagree with the generic definition of Buchanan and Gibbons (1974). Produces a non-diffusible reddish brown or pink pigment (Long and Hammer, 1941).

Nitrates reduced to nitrites, but further reduction variable (King and Phillips, 1978). Able to grow at 4 °C but not at 37 °C and produces phosphatase rapidly in milk and other media (Long and Hammer, 1941). Possesses ornithine decarboxylase (King and Phillips, 1978). Produces $H_2S$ in peptone iron agar (Levin, 1968).

Recorded as facultative (Long and Hammer, 1941), but probably due to ability to use nitrate as terminal electron acceptor. Optimum growth temperature is approximately 21 °C (Derby and Hammer, 1931).

**The Genus *Brucella***

Members prone to spontaneous mutation. May produce a number of colony types: smooth (S) phase (most common), rough (R), intermediate (I) and mucoid (M). If a culture normally produces S phase colonies, then only S or I types should be used for serological identification. Non-motile. Requirement for added $CO_2$ should be determined on primary isolation as dependence may fade. $H_2S$ production is altered by medium composition and, therefore, reference strains should be included as controls.

(*i*) *Brucella abortus*

Usually pathogenic for cattle causing abortion, and can also infect man. Non-pigmented, circular convex colonies are produced with a smooth glistening surface and an entire edge on the enriched medium described below.

Oxidase-positive. Nitrates reduced to nitrites. Usually require added $CO_2$ (5%) for growth, and produce $H_2S$. Have abortus antigen predominant. Nine biotypes recognised, subdivided mainly on serology, ability to produce $H_2S$ in suitable media and $CO_2$ requirement.

Most strains grow poorly on peptone medium, growth can be improved by the use of tryptose and trypticase, particularly if supplemented with 1–5% v/v serum. Growth temperature range is 20–40 °C, with optimum at 37 °C.

(*ii*) *Brucella melitensis*

Usually pathogenic for goats and sheep, but can infect other species including cattle and man. Colonial appearance is similar to *Brucella abortus*. Oxidase-positive. Nitrates reduced to nitrites. Produces little or no $H_2S$ on peptone media. Usually melitensis antigen predominant. Three biotypes recognised, differentiated mainly on serology.

Growth temperature range is 20–40 °C, with optimum at 37 °C.

## FAMILY ENTEROBACTERIACEAE

This is a Family of Gram-negative, facultatively anaerobic rods. There are twelve genera listed as belonging to this Family in the 8th edition of

*Bergey's Manual* (Buchanan and Gibbons, 1974) and all of them (with the exception of *Erwinia* and possibly *Edwardsiella*) could be considered to have potential associations with milk. The genera comprising the family are: *Escherichia*, *Edwardsiella*, *Citrobacter*, *Salmonella*, *Shigella*, *Klebsiella*, *Enterobacter*, *Hafnia*, *Serratia*, *Proteus*, *Yersinia* and *Erwinia*.

**General Characteristics of All Genera**

Cells rod-shaped, straight and small (0·4–0·7 × 1·0–4·0 μm). Usually occur singly. May be capsulated. Motile by means of peritrichous flagella or non-motile. Gram-negative. Aerobic and facultatively anaerobic. Chemo-organotrophs. The optimum growth temperature of all genera except *Erwinia* is 37 °C, although the optimum for *Yersinia* may be closer to 30 °C. Some species live in the intestines of man and other animals sometimes causing intestinal disturbances. Some are plant pathogens (*Erwinia* spp. which are not discussed here) and others are saprophytic causing decomposition of dead organic materials. None are particularly heat resistant and, thus, all are easily eliminated from milk by pasteurisation treatments.

Colonies of the Enterobacteriaceae are morphologically homogeneous on nutrient agar, being usually smooth, low convex, moist with a shiny surface, having an entire edge and are grey/white. Genera cannot be distinguished from one another on colony morphology.

The main biochemical characteristics of all genera except *Erwinia* are given in Table I (Buchanan and Gibbons, 1974).

In relation to this Family it is necessary to explain the term 'coliform', since it is still found extensively in the literature. Coliforms may be defined as Gram-negative, oxidase-negative, non-sporing rods which can grow aerobically on an agar medium containing bile salts and ferment lactose within 48 h at 37 °C with the production of both acid and gas (Department of Health and Social Security, 1969). It includes organisms belonging to several genera within the Family, e.g. *Escherichia*, *Enterobacter*, *Citrobacter*, etc. In the presumptive test for coliforms employed in the dairying field, a positive result depends on the production of acid and gas in MacConkey's broth after 72 h incubation at 30 °C (British Standards Institution, 1968).

Tests for the presence of coliform organisms are of limited value for raw milk and cream, etc., since they gain easy access from the intestinal tract of the cow. However, a coliform and phosphatase test are often carried out on pasteurised milk, cream and dairy products where the presence of coliforms, together with a positive phosphatase test, indicates improper

TABLE I

MAIN BIOCHEMICAL CHARACTERISTICS OF GENERA BELONGING TO ENTEROBACTERIACEAE (EXCEPT *Erwinia*)

| *Characteristic* | Escherichia | Edwardsiella | Citrobacter | Salmonella | Shigella | Klebsiella | Enterobacter | Hafnia | Serratia | Proteus | Yersinia |
|---|---|---|---|---|---|---|---|---|---|---|---|
| Catalase | + | + | + | + | D | + | + | + | + | + | + |
| Oxidase | – | – | – | – | – | – | – | – | – | – | – |
| β-galactosidase | + | – | + | D | d | + | + | + | + | – | + |
| Gas from glucose at 37 °C | + | + | + | + | – | d | + | + | d | D | – |
| Nitrate reduction | + | + | + | + | + | + | + | + | + | + | + |
| Lactose (acid from) | + or × | – | + or × | D | D | D | + | – | – | – | – |
| Citrate utilisation as sole C source | – | – | + | + | – | d | + | + | + |  | – |
| Methyl red test | + | + | + | + | + | D | – | – | – | + | + |
| Voges–Proskauer test | – | – | – | – | – | D | + | + | D | d | – |
| Indole production | + | + | D | – | D | d | – | – | – | D | D |
| $H_2S$ from TSI† | – | + | D | + | – | – | – | – | – | D | D |
| Lysine decarboxylated | + | + | – | + | – | d | D | + | + | d | – |
| Urea hydrolysed | – | – | (+) | – | – | d | (d) | – | – | D | D |
| Phenylalanine deamination | – | – | – | – | – | – | – | – | – | + | – |

D = Different reactions given by different species of a genus.
d = Different reactions given by different strains of a species or serotype.
× = Late and irregularly positive (mutative).
( ) = Delayed reaction.
† Triple sugar iron agar.

pasteurisation, whereas the presence of coliforms together with a negative phosphatase test indicates post-pasteurisation contamination.

The following four genera are worthy of further consideration.

## The Genus *Escherichia*

This genus contains only one species, *E. coli*. Glucose and other carbohydrates are fermented by means of the mixed/formic acid fermentation to give lactic, acetic and formic acid. Part of the formic acid is split, under acid conditions, by a formic hydrogenlyase enzyme system into equal amounts of $CO_2$ and $H_2$. The organism may also produce gas when involved in spoilage incidents. This mixed-acid fermentation enables the organism to be differentiated from other coliforms, by means of the group of tests known collectively as the IMViC reactions, i.e. indole production, Methyl Red, Voges–Proskauer and citrate utilisation tests. *Escherichia coli*, if subjected to the IMViC tests, would give (+ + − −) results respectively, whereas another coliform, *Enterobacter aerogenes*, which carries out the butanediol–formic fermentation would give (− − + +) results. Part of this information is made use of in the microbiological analysis of drinking water or water for food processing.

Some strains produce toxins and/or specific antibiotic proteins (colicins), both of which are plasmid determined (Buchanan and Gibbons, 1974). *Escherichia coli* is found in the lower part of the intestine of most warm blooded animals, and consequently contaminates milk either directly or indirectly via faecal material. Some strains may be opportunistic pathogens, and an example of this behaviour is the ability of some strains to cause acute mastitis. The use of deep sawdust bedding, which rapidly becomes wet and soiled with faecal material, is thought to exacerbate the problem.

*Escherichia coli* under suitable conditions can spoil milk and most dairy products, usually with the production of gas and an 'unclean' or faecal smell. It has been isolated from spoiled cans of evaporated and sweetened condensed milk (Crossley, 1946), and can cause 'early blowing' of various types of cheese (Davis, 1965). Some encapsulated strains of *E. coli* can cause ropiness in milk, as can some strains of *Citrobacter* and *Klebsiella*, even when milk is stored at low temperatures (Thomas, 1958).

**The Genus *Salmonella***

Special growth factors are not required and most strains will grow on defined media. Most strains produce gas. However, *S. typhi*, the causative organism of typhoid fever, never produces gas.

In 1968 there were more than 1200 closely related types (serotypes rather than species) of salmonellae classified into serotypes by means of Kauffmann–White scheme, in which numbers and letters are given to the principal O (somatic), Vi (virulence) and H (flagellar) antigens. Many strains are pathogenic for man and/or animals by means of the endotoxin produced. The disease may take the form of an enteric fever (septicaemia) as with *S. typhi* or, more commonly, an enteritis as with *S. typhimurium* (Cruickshank *et al.*, 1973). This organism may also contaminate milk via faecal material, although cows suffering from salmonellosis can, under certain conditions, excrete viable organisms in their milk (Hobbs and Gilbert, 1978). Salmonellae have, in the past, been responsible for relatively serious food poisoning outbreaks through the consumption of raw milk; *S. dublin* and *S. typhimurium* have been implicated (Dewberry, 1959).

### The Genus *Enterobacter*

Some strains are encapsulated. There are two species, *Ent. cloacae* and *Ent. aerogenes*. As stated earlier, *Ent. aerogenes* ferments glucose by means of the butanediol–formic fermentation. A positive Voges–Proskauer test is thus usually obtained.

Organisms belonging to this genus may be of faecal origin, or from soil, vegetation, water, etc., and they may contaminate milk from any of these sources. Some encapsulated strains of *Ent. aerogenes* have the ability to cause ropiness in milk (Seaman, 1963).

### The Genus *Yersinia*

There are three species: *Y. pestis*, *Y. pseudotuberculosis* and *Y. enterocolitica*. All are non-motile at 37 °C, but two are motile at temperatures below 37 °C. No encapsulation is found.

All species are pathogenic for man and animals. *Yersinia enterocolitica* is a psychrotrophic organism. It contaminates food and milk either directly or indirectly via faeces, urine or insects (Stern and Pierson, 1979). Milk has been implicated as the source of two outbreaks of gastro-enteritis caused by *Y. enterocolitica* in Canada (Hughes, 1979), and certainly the organism has been isolated from raw and pasteurised milk in various parts of the world, although most strains were considered to be non-pathogenic for humans (Stern and Pierson, 1979).

Although these organisms grow best at 30–37 °C, their growth temperature range is −2 °C to 45 °C (Buchanan and Gibbons, 1974).

## FAMILY VIBRIONACEAE

There are five genera belonging to this Family and nine genera of uncertain affiliation associated with it. Only one of these true genera, *Aeromonas*, and two genera of uncertain affiliation, *Flavobacterium* and *Chromobacterium*, could be considered to be associated with milk. However, organisms belonging to another of the true genera, *Vibrio*, have been isolated from butter (Lightbody and Petersen, 1962). The other true genera are *Plesiomonas*, *Photobacterium* and *Lucibacterium* and the other genera of uncertain affiliation are *Zymomonas*, *Haemophilus*, *Pasteurella*, *Actinobacillus*, *Cardiobacterium*, *Streptobacillus* and *Calymmatobacterium*.

**General Characteristics of All Genera**

Cells rod-shaped, straight or curved (0·2–2·0 × 0·5–6·0 μm). Usually motile by means of polar flagella. Gram-negative. Aerobic and facultatively anaerobic. Chemo-organotrophs. Oxidase-positive. May be mesophilic or psychrotrophic. Most genera are of water or soil origin.

**Further Comments on the Genera *Aeromonas*, *Chromobacterium* and *Flavobacterium***

These genera may be differentiated from each other by means of colony pigmentation. Colonies of organisms belonging to the genus *Aeromonas* are generally non-pigmented with the exception of *A. salmonicida*, which produces a brown water-soluble pigment. However, this organism is unlikely to be isolated from milk or dairy products. Colonies of organisms belonging to the genus *Chromobacterium* are pigmented violet, due to the production of the water-insoluble violacein, and those belonging to the genus *Flavobacterium* are pigmented yellow, orange, red or brown, probably due to the production of water-insoluble carotenoid pigments (Buchanan and Gibbons, 1974).

Organisms belonging to the genus *Aeromonas* and *Chromobacterium* are generally not fastidious. However, organisms belonging to the genus *Flavobacterium* are comparatively fastidious because of nitrogen requirements and exogenous requirements for B complex vitamins.

Most organisms belonging to any of the three genera grow best at 25–30 °C, although they differ in their maximum and minimum growth temperatures. Many will grow on plates incubated for the isolation or counting of psychrotrophs in milk and dairy products, and organisms belonging to each of the three genera viz., *Aeromonas*, *Flavobacterium* and *Chromobacterium* have been reported as having been isolated from these sources by Dempster (1968), Thomas (1958) and Witter (1961), respectively. However, although organisms belonging to the genus *Chromobacterium* are commonly found in soil and water, they are comparatively infrequent in milk (Thomas, 1958).

It is possible that some strains of organisms belonging to these genera are producers of partially heat stable extracellular enzymes and may, therefore, contribute to the psychrotrophic spoilage of milk and dairy products. In addition, a *Flavobacterium* sp. has been reported as causing ropiness in milk at a temperature of 4 °C (Thomas, 1958).

## FAMILY NEISSERIACEAE

This Family comprises four genera, viz., *Neisseria*, *Branhamella*, *Moraxella* and *Acinetobacter*. There are also two genera of uncertain affiliation: the genus *Paracoccus* and the genus *Lampropedia*. Only the genus *Acinetobacter* and organisms described as being *Moraxella*-like have been found associated with water and food including milk and milk products.

### General Characteristics of All Genera

Cells plump, rod-shaped (1·0–1·5 × 1·5 × 2·5 μm) occurring in pairs or short chains, or may be spherical occurring in pairs or in masses with adjacent sides flattened. Non-motile. Gram-negative. Aerobic. Chemo-organotrophs. Optimum growth temperature 32–37 °C, but some are able to grow at temperatures much lower than this.

### Further Comments on the Genus *Acinetobacter* and on Organisms Classified as being *Moraxella*-like

*Moraxella*-like organisms may be differentiated from organisms belonging to the genus *Acinetobacter* by means of the oxidase test. *Moraxella*-like organisms are oxidase-positive, whereas *Acinetobacter* are oxidase-negative.

The *Moraxella* organisms are fastidious and are pathogens of warm blooded animals including man. The *Moraxella*-like organisms of interest here, besides being oxidase-positive, conform to the morphology of *Moraxella*. They have been isolated from foods including milk stored under cold conditions (Lautrop, unpublished; Neill, 1974). These organisms may or may not grow at 37 °C. Their taxonomic position at the present time is uncertain (Buchanan and Gibbons, 1974).

Organisms belonging to the genus *Acinetobacter* are not fastidious and are saprophytic. They will grow well at 30–32 °C, and have been isolated from milk both as psychrotrophs and as mesophiles (Thomas and Thomas, 1973; Neill, 1974). *Acinetobacter viscolactis* is thought to be one of the causative organisms of ropiness in milk, although not all strains are capable of producing the capsule responsible for ropiness. As part of the psychrotrophic flora of raw milk, it is possible that some strains of both these genera also produce extracellular degradative enzymes.

## FAMILY MICROCOCCACEAE

This Family comprises three genera: *Micrococcus*, *Staphylococcus* and *Planococcus*; of which the first two are considered worthy of further description.

## General Characteristics of All Genera

Cells spherical 0·5–3·5 μm in diameter. Gram-positive. Usually non-motile. Catalase-positive. Chemo-organotrophs. Metabolism respiratory or fermentative. Aerobic or facultatively anaerobic.

The genus *Micrococcus* is usually distinguished from the genus *Staphylococcus* by the latter's ability to ferment glucose (oxidation/fermentation test described by Baird-Parker, 1963). However, with this scheme, it is difficult to separate coagulase-negative, weakly anaerobic staphylococci from micrococci, and such strains are often misidentified as micrococci. This can be resolved by using the spot test for lysostaphin sensitivity (Schleifer and Kloos, 1975) which relies on differences in cell wall chemistry brought about by differing guanine plus cytosine (G + C) contents of their DNA. All the coagulase-negative staphylococci (G + C: 30–38 mol %) tested by these authors showed some lysostaphin sensitivity, while most micrococci (G + C: 66–73 mol %) tested were resistant.

## The Genus *Micrococcus*

### (*i*) *Micrococcus luteus*

Common in soil, water and on skin of man and animals. Not pathogenic. Colonies are yellow, yellowish green or orange. Some strains produce a violet water-soluble pigment.

Usually no detectable acid from glucose. Sensitive to novobiocin (minimum inhibitory concentration (MIC) $< 1\ \mu g\ ml^{-1}$) (Mitchell and Baird-Parker, 1967; Jeffries, 1969). Nitrates usually not reduced.

Some strains require a complex medium. Strict aerobe. Most strains grow at 10 °C and some at 45 °C. Optimum 30 °C.

### (*ii*) *Micrococcus varians*

Common in milk and dairy products, animal carcasses and soil. Not pathogenic. Colonies are yellow, smooth and convex with regular edge.

Resistant to novobiocin (MIC $> 2{\cdot}0\ \mu g\ ml^{-1}$) (Mitchell and Baird-Parker, 1967; Jeffries, 1969). Nitrates and nitrites usually reduced.

Growth reported only on complex media. Strict aerobe. Growth by most strains at 10 °C but not at 45 °C. Optimum 22–37 °C.

## The Genus *Staphylococcus*

### (*i*) *Staphylococcus aureus*

Found chiefly in nasal membranes and on the skin of animals and man. Potential pathogen causing a wide range of infections and food-borne intoxications. Colonies of most strains are orange, although certain antibiotic-resistant and bovine strains are more frequently yellow. Pigment production depends on growth conditions, and may be variable within a single strain. At least three haemolysins are produced (alpha, beta and delta) (Elek and Levy, 1950). Produces heat-resistant endonucleases. Differentiation of strains by phage typing is common. May produce coagulase.

Food poisoning potential is due to the elaboration of a heat-stable enterotoxin (survives boiling for 30 min). Symptoms of nausea, vomiting and diarrhoea are most common 2–6 h after ingestion of enterotoxin. Recovery 24–48 h.

Growth requirements complex. Facultative anaerobe. Most strains grow between 6·5 and 46 °C, optimum 30–37 °C.

(*ii*) *Staphylococcus epidermidis*

Common in nature, but most frequently found on skin and mucous membranes of man and animals. Many strains may be primary pathogens or secondary invaders, but some show a commensal relationship. Colonies are circular, convex with a smooth or slightly granular surface and an entire or slightly irregular edge. Usually the colonies are white or yellow pigmented, occasionally orange and rarely purple.

Produces no coagulase. Weak haemolytic activity by some strains. Failure to ferment trehalose and mannitol is characteristic of this species.

Complex growth requirements. Facultative anaerobe. Grows at 45 °C and often at 10 °C, optimum 30–37 °C.

(*iii*) *Staphylococcus saprophyticus*

Commonly isolated from air, soil, dairy products and the surface of animal carcases. Usually regarded as non-pathogenic. Colonies are smooth, convex with a regular or slightly irregular margin. They are usually white but may be occasionally yellow or orange.

Strains characteristically do not reduce nitrate, but most produce acetylmethylcarbinol and ferment xylitol (Schleifer and Kloos, 1975).

Growth reported only on complex media. Facultative anaerobe. Most strains grow at 10 °C and some at 45 °C, optimum 30–37 °C.

## FAMILY STREPTOCOCCACEAE

The genera *Streptococcus*, *Leuconostoc*, *Pediococcus*, *Aerococcus* and *Gemella* are included in this Family. Only *Streptococcus* and *Leuconostoc* will be considered further.

## General Characteristics of All Genera

Cells spherical or ovoid and arranged in pairs, tetrads or chains of varying length. Gram-positive. Rarely motile. Catalase test usually negative, benzidine test negative. Chemo-organotrophs. Metabolism fermentative. Facultatively anaerobic.

## The Genus *Streptococcus*

Cells divide in one plane. Glucose fermented yielding chiefly dextrorotatory lactic acid—homofermentative. Catalase-negative.

This genus was divided into serological groups by Lancefield (1933). Sherman (1937) later divided the streptococci into four primary divisions as shown in Table II. For a review on the composition and differentiation of this genus see Jones (1978).

### (*i*) *Streptococcus pyogenes*

Found in the human mouth, throat, respiratory tract, blood, various lesions and inflammatory exudates. Causative organism of scarlet fever. Colonies are non-pigmented, mucoid, glossy or matt.

Pyogenic group. Lancefield's group A. Most strains produce a strong $\beta$-haemolysis on blood agar within 24 h. The majority of strains elaborate an enzyme, fibrinolysin, which activates a protease present in normal human plasma causing lysis of clots.

Complex growth requirements. Optimum growth temperature 37 °C.

### (*ii*) *Streptococcus equisimilis*

Has been isolated from the upper respiratory tract of normal and diseased humans and animals. It has occasionally been associated with erysipelas and puerperal fever. Colonial appearance similar to *Str. pyogenes.*

Pyogenic group. Lancefield's group C. Most strains are fibrinolytic, with the exception of animal strains which may not lyse human fibrin.

Growth reported only on complex media. Optimum growth temperature 37 °C.

### (*iii*) *Streptococcus zooepidemicus*

Found in blood, inflammatory exudates and lesions of diseased animals. Can cause septicaemia in cows and swine. Not pathogenic for humans. Colonial appearance similar to *Str. pyogenes.*

Pyogenic group. Lancefield's group C. The production of acid from

TABLE II
DIVISION OF THE GENUS *Streptococcus*

| *Sherman's group* | *Lancefield's group* | *Haemolysis* | *Growth at* | | *Presence of growth in* | | *0·1% Methylene blue* | *Survives 60°C for 30 min* | *$NH_3$ from arginine* |
|---|---|---|---|---|---|---|---|---|---|
| | | | *10°C* | *45°C* | *pH 9·6 broth* | *6·5% NaCl broth* | | | |
| Pyogenic | ABCEFGH | $\beta$ | − | − | − | − | − | − | + |
| Viridans | No group specific antigen demonstrated | $\alpha$ or $\gamma$ | − | + | − | − | − | ± | − |
| Lactic | N | $\gamma$ | + | − | − | − | + | ± | ± |
| Enterococcus | D | $\alpha$, $\beta$ or $\gamma$ | + | + | + | + | + | + | + |

sorbitol, but not from trehalose, and its inability to produce fibrinolysin is characteristic of this species.

Complex growth requirements. Optimum growth temperature 37 °C.

*(iv) Streptococcus dysgalactiae*

Has been found in milk and udders of cows with mastitis. Colonial appearance similar to *Str. pyogenes.*

Pyogenic group. Lancefield's group C. May produce a fibrinolysin active against bovine fibrin but not human fibrin. Edward's aesculin Crystal Violet blood agar (Edwards, 1933) is selective for mastitic streptococci. Aesculin fermenting *Str. uberis* produces black colonies, non-aesculin fermenting *Str. dysgalactiae* and *Str. agalactiae* produce colourless colonies.

Complex growth requirements. Optimum growth temperature 37 °C.

*(v) Streptococcus agalactiae*

Has been isolated from milk and udder tissues of mastitic cows. Some strains produce colonies showing a yellow, orange or brick-red pigmentation.

Pyogenic group. Lancefield's group B. Weak cross reactions may occur between groups B and G. Approximately half the bovine strains produce a narrow zone of $\beta$-haemolysis, while other strains may show $\alpha$-, $\beta$- or $\gamma$-reactions. Selective media—see previous section on *Str. dysgalactiae.*

Complex growth requirements. Optimum growth temperature 37 °C.

*(vi) Streptococcus acidominimus*

Commonly found in bovine vagina and occasionally on the skin of calves and in raw milk.

Pyogenic group. $\gamma$-haemolysis. Characterised by the inability of most strains to reduce the pH value of carbohydrate-containing media below pH 6·0. Can be distinguished from *Str. agalactiae* by its inability to ferment glycerol or hydrolyse arginine.

Growth only on complex media. Optimum growth temperature 37 °C.

*(vii) Streptococcus bovis*

Isolated from the alimentary tract of cows, sheep and other ruminants, and has been found in large numbers in human faeces.

Enterococcus group. Lancefield's group D. Cross reactions can occur with groups E and N. Most strains produce $\alpha$-haemolysis though some may give a $\gamma$-reaction on blood agar. Can be distinguished from *Str. equinus* by the latter's inability to produce acid from lactose.

Not auxotrophic for any specific amino acid and is the least nutritionally fastidious *Streptococcus* sp. Optimum growth temperature 37 °C.

*(viii) Streptoccocus thermophilus*

Thermophilic starter, used in combination with other starters for the production of yoghurt, Swiss-type and Italian cheeses. May be used for the detection of inhibitory substances in milk.

Viridans group. γ-reaction on blood agar. Characterised by its maximum growth temperature (grows at 50 °C but not at 53 °C) and its heat resistance (survives 65 °C for 30 min).

Complex nutritional requirements. Optimum growth temperature 40–50 °C.

*(ix) Streptococcus faecalis*

Isolated from intestines of man and animals. Its presence is often correlated with faecal contamination, e.g. in bacteriological water analysis. It is considered to be more resistant to freezing, low pH, moderate heat treatment and marginal chlorination than *Escherichia coli* (Harrigan and McCance, 1966).

Colonies appear smooth, entire and are rarely pigmented (false pigmentation may be due to the precipitation of metal ions (Jones *et al.*, 1963)).

Enterococcus group. Lancefield's group D. Characteristically grows in the presence of 0·04% tellurite (Skadhauge, 1950). Selective media—Mead's medium, Barne's medium, maltose azide broth, maltose azide tetrazolium agar and m-Enterococcus agar.

Complex nutritional requirements. Optimum growth temperature 37 °C.

*(x) Streptococcus uberis*

Found in cow's milk, and in the throat and faeces of cows. Responsible for a form of bovine mastitis which is particularly common in winter. Serology of this species is confused, reacts with Lancefield's groups E, P, U, C, G and D antisera (Roguinsky, 1969; 1971). Weak α- and γ-reaction on blood agar.

Complex nutritional requirements. Optimum growth temperature 35–37 °C.

*(xi) Streptococcus lactis*

Mesophilic starter, used alone or in combination with other starters in the production of hard pressed cheese, e.g. Cheddar, Gouda, and many other types. Sole starter in Taette milk.

Lactic group. Lancefield's group N. A weak α- or γ-reaction on blood agar. Ammonia produced from arginine, acid from maltose, but no $CO_2$ or di-acetyl produced from citrate. Some strains may metabolise leucine to produce 3-methyl-butanol which gives a malty flavour defect in dairy products. May produce nisin, an antimicrobial compound active against Gram-positive organisms, e.g. *Staph. aureus*, spore formers, streptococci and lactobacilli.

Complex nutritional requirements. Optimum growth temperature is approximately 30 °C.

*(xii) Streptococcus lactis sub-sp. diacetylactis*

Mesophilic starter, used in combination with other starters to produce, hard pressed cheese, mould ripened and soft ripened cheese, cottage cheese and cream cheese, cultured butter, buttermilk, Quarg and many other products.

Lactic group. Lancefield's group N. Possesses same characteristics as *Str. lactis*, except that it is capable of producing $CO_2$ and diacetyl from citrate. Also capable of producing acetic acid from citrate, which is reported to suppress pseudomonads, coliforms and salmonellae (Babel, 1977).

Complex growth requirements. Optimum growth temperature is approximately 30 °C.

*(xiii) Streptococcus cremoris*

Mesophilic starter, used in combination with other starters in the production of hard pressed cheeses, mould ripened and soft ripened cheeses, Feta and many other cheeses.

Lactic group. Lancefield's group N. A weak α- or γ-reaction on blood agar. No ammonia produced from arginine, no acid from maltose, no $CO_2$ or diacetyl produced from citrate.

Complex growth requirements. Optimum growth temperature is approximately 30 °C.

**The Genus *Leuconostoc***

Glucose fermented with production of D(−) lactic acid, ethanol and $CO_2$, i.e. heterofermentative.

*(i) Leuconostoc mesenteroides*

Found in milk and dairy products, slimy sugar solutions and on fruit and vegetables. Dextran produced by some strains may be used as a stabiliser in ice-cream mixes.

A characteristic slime of dextran is usually produced from sucrose and is favoured by a temperature of 20–25 °C. Some strains, particularly those from dairy sources, produce little dextran. Acid produced from sucrose and trehalose, and usually from arabinose. Does not withstand heating to 55 °C for 30 min, although slimy cultures in sugar solutions may survive 80–85 °C.

The number of amino acids essential for growth is small, only valine and glutamic acid being required by all strains. Growth temperature range 10–37 °C. Optimum 20–30 °C.

*(ii) Leuconostoc dextranicum*

Found in milk and milk products and on fruit and vegetables.

Dextran is formed from sucrose, but not as actively as with most strains of *L. mesenteroides*. Acid from sucrose and trehalose but not arabinose.

Amino acid requirements are more varied than with *L. mesenteroides*, but only a few amino acids affect the growth of any one strain. Temperature range 10–37 °C. Optimum 20–30 °C.

*(iii) Leuconostoc paramesenteroides*

Found in milk and milk products, fermenting vegetables and on herbage. Apparently widely distributed.

No dextran formation. Acid produced from sucrose and trehalose, and usually from arabinose. Tolerant of higher concentrations of NaCl (6·5 %) than other species.

Complex nutritional requirements. Growth temperature range 10–37 °C. Optimum 20–30 °C.

*(iv) Leuconostoc lactis*

Found in milk and dairy products.

No dextran production. Acid from sucrose, but not from arabinose or trehalose. More heat resistant than other species, and normally survives heating at 60 °C for 30 min.

Complex nutritional requirements. Growth temperature range 10–40 °C. Optimum 25–30 °C.

*(v) Leuconostoc cremoris*

Mesophilic starter and used in combination with other starters in the production of cottage and cream cheese, cultured butter, buttermilk and Quarg.

No dextran production. No acid produced from arabinose, sucrose or trehalose. Produces acetic acid from citrate (see section on *Str. lactis* subsp. *diacetylactis*), and may be added to cottage cheese to prevent spoilage by slime-producing pseudomonads (Babel, 1977). Characterised by limited fermentation ability and complex growth factor requirements.

Growth temperature range 10–30 °C. Optimum 18–25 °C.

## FAMILY BACILLACEAE

There are five genera belonging to this Family of which two are of primary importance to the dairy microbiologist, viz., *Bacillus* and *Clostridium*. Other genera in this Family are *Sporolactobacillus*, *Desulfotomaculum*, *Sporosarcina* and *Oscillospira*, which is a genus of uncertain affiliation.

### General Characteristics of the Genera *Bacillus* and *Clostridium*

Cells rod-shaped, generally straight and may be fairly large (0·3–2·2 × 1·2–7·0 μm). May occur singly, in pairs or in chains. Endospores formed which are resistant to heat and other destructive agents. Motile by lateral or peritrichous flagella or non-motile. Gram-positive except in later stages of growth. Aerobic, facultative or anaerobic. Chemo-organotrophs.

Organisms assigned to the genus *Bacillus* aerobically form refractile spores, while *Clostridium* spp. form spores under strictly anaerobic conditions. In addition, most species of *Bacillus* form catalase, whereas *Clostridium* spp. do not. The production of catalase also distinguishes *Bacillus* spp. from the micro-aerophilic *Sporolactobacillus* spp.

### The Genus *Bacillus*

Typical habitat is soil, although they are distributed widely in nature and may gain access to milk and dairy products through various routes, e.g. air, water, fodder and feed. Mikolajcik (1978) has suggested that there may be a potential spoilage problem in milk due to the growth of psychrotrophic spore-formers.

(*i*) *Bacillus cereus*

*B. cereus* is important in the dairy and food industries on two main counts.

1. Toxin production; it has been proved to be the causative organism of food poisoning due to the ingestion of contaminated fried rice.

Numbers of organisms required to cause poisoning in this situation are, however, in excess of $10^6$ $g^{-1}$, and the problem is not thought to exist in the dairy sector due to comparatively low numbers (Hobbs and Gilbert, 1978).

2. Enzyme production; extracellular protease and phospholipase C (lecithinase) production, combined with the fact that heat-resistant spores are produced, results in the organism giving rise to sweet curdling and bitty cream in pasteurised milk (Stewart, 1975). Exceptionally heat-resistant spores of *B. cereus* present a potential spoilage problem in ultra-high temperature (UHT) milk products (Burton, 1977).

Colonial appearance varies greatly between different strains, e.g. may have dull or frosted glass appearance with an undulate margin, or the colony may form root-like outgrowths which spread widely.

*B. cereus* strains have a requirement for one or more amino acids (varies according to strain). Vitamins are not required. The organism will grow readily at 20–35 °C, but will not grow below 10 °C.

(*ii*) *Bacillus subtilis*

Jayne-Williams and Franklin (1960) included this organism, alongside *B. cereus*, as being the two species most commonly encountered in milk. *B. subtilis* has been implicated in causing ropiness or sliminess in raw and pasteurised milk, and has also been reported to be a causative organism in the sweet curdling of milk (Jayne-Williams and Franklin, 1960).

Colony form varies from round to irregular. Surface dull, and may be wrinkled, becoming cream-coloured or brown. Appearance varies greatly with medium composition.

Extracellular products include enzymes which will degrade pectin, polysaccharides of plant tissues and casein. Levan is formed extracellularly from sucrose and raffinose, and polypeptide antibiotics are produced.

No vitamins or amino acids are required for vegetative growth. The organism will grow readily at 20–45 °C.

(*iii*) *Bacillus stearothermophilus*

While this organism, being thermophilic, is not of major importance in the liquid milk industry, it is, however, worthy of consideration here due to its potential for spoiling canned foods including dairy products. Anatskaya and Efimova (1978) found *B. stearothermophilus* to be implicated in the spoilage of evaporated sterilised milk. The source of the organism is probably the rice grains, starch or sugar used in canned dairy products rather than from the milk itself. In general, the spores are more heat resistant than those of mesophilic organisms, and spores surviving canning treatments can cause flat sour spoilage (no gas production and, therefore, no distension of can ends occurs).

Colonies vary from round to irregular, translucent to opaque and from smooth to rough. They are not distinctive and are small to pinpoint in size.

The most distinctive diagnostic characters are its ability to grow at 65 °C, and its limited tolerance to acid (will not grow in food if pH value $< 5$).

Minimal nutritional requirements vary greatly, and range from a carbon source and $NH_4$-N in the absence of growth factors, to requirements for various amino acids and vitamins. The organism will grow between 45–65 °C, with a maximum growth temperature of 75 °C and a minimum of 30 °C.

*(iv) Bacillus coagulans*

This organism has been implicated in the spoilage (coagulation) of canned evaporated milk (Foster *et al.*, 1958: Anatskaya and Efimova, 1978). Perhaps the source of the organism is silage in which the organism may multiply.

Colonies are usually small, round and opaque.

The organism is aciduric and, although for initiation of growth a pH value of about 6 is required, some strains will grow as low as pH 4·0. The main product of glucose fermentation is L(+) lactic acid.

Minimal nutritional requirements are diverse, but certain vitamins and amino acids are usually required. The organism will grow readily between 25–55 °C.

**The Genus *Clostridium***

These organisms are found in sediments of various types, and in the intestinal tract of man and animals. They gain entry to milk via faeces, soil and feedingstuffs, especially silage (Stewart, 1978). Since most strains are strictly anaerobic, they are of greatest potential importance in the area of cheese production (Davis, 1965) and canned milk products (Hersom and Hulland, 1969).

*(i) Clostridium thermosaccharolyticum*

This organism has been isolated from sterilised milk (Candy and Nichols, 1956) although, due to its thermophilic and anaerobic characteristics, it is only likely to grow and cause spoilage under exceptional circumstances.

Surface colonies are circular, low and flat with a raised centre, translucent greyish and with a glossy surface.

This organism belongs to Group III in *Bergey's Manual* (Buchanan and Gibbons, 1974) viz., the spore position is terminal and gelatin is not hydrolysed. Saccharolytic fermentation products include acetic and butyric acids. Lactose is fermented and milk is coagulated.

The optimum temperature for growth is 55 °C.

*(ii) Clostridium butyricum*

This organism has been implicated as a causative organism of 'late blowing' of cheese, although other species (described below) are undoubtedly also involved. This problem of 'late blowing' (extensive gas production 1–2 months after manufacture) was found to be associated with the feeding of heavily contaminated silage. The problem is controlled by various factors viz., numbers of spores in the milk, pH value, salt concentration and degree of anaerobiosis in the cheese. Certain types of cheese are consequently more susceptible to this problem than others. Emmenthal and Gruyère are especially susceptible (Davis, 1965).

Surface colonies are circular, medium sized, slightly raised and white to cream in colour.

This organism belongs to Group I in *Bergey's Manual* (Buchanan and Gibbons, 1974) viz., spore position is subterminal and gelatin is not hydrolysed. Saccharolytic fermentation products include acetic acid, butyric acid and butanol. Lactose is fermented and milk becomes acid with early coagulation; the fermentation is often stormy. No digestion of the clot occurs.

Grows best at 25–37 °C.

*(iii) Clostridium tyrobutyricum*

This organism has been isolated from rancid Finnish Emmenthal cheese (Korhonen *et al.*, 1978), and has also been implicated as a causative organism of late (butyric) blowing of cheese made with silage milk (Bergère, 1978), and of blowing of Grana cheese (Matteuzzi *et al.*, 1977). The organism is able to multiply in cheese (and thus causes blowing) due chiefly to its resistance to acid pH values and salt (Bergère and Hermier, 1970).

Surface colonies are small, circular, convex, grey and translucent.

Like *Cl. butyricum*, this organism also belongs to Group I in *Bergey's Manual* (Buchanan and Gibbons, 1974). Saccharolytic fermentation products include large amounts of acetic and butyric acids.

Optimum temperature for growth is 37 °C.

(*iv*) *Clostridium sporogenes*

This organism was also considered to be of importance in the development of rancidity of Finnish Emmenthal cheese (Korhonen *et al.*, 1978).

Surface colonies are large, raised, white to yellow centre with grey rhizoids, 'Medusa head' margin, semi-opaque, with a matt surface.

Belongs to Group II in *Bergey's Manual* (Buchanan and Gibbons, 1974) viz., spore position subterminal and gelatin is hydrolysed. Proteolytic, but acid production from carbohydrates is often masked by ammonia from amino acid deamination. Butyric acid is the main fermentation product. Milk is digested.

Grows best at 30–40 °C.

## FAMILY LACTOBACILLACEAE

This Family comprises one true genus, *Lactobacillus*, together with three genera of uncertain affiliation, viz. *Listeria*, *Erysipelothrix* and *Caryophanon*. Only the true genus and the genus *Listeria* are worthy of consideration here.

### General Characteristics of All Genera

Straight or curved rods. Gram-positive. Rarely motile. Catalase-negative (rare strains decompose peroxide by a pseudocatalase). Benzidine reaction negative. Anaerobic or facultative.

### The Genus *Lactobacillus*

This genus was subdivided by Orla Jensen (1919; 1943) into three groups as shown in Table III.

Surface growth of these organisms is enhanced on enriched media and under anaerobic conditions with added $CO_2$ (5–10 %).

TABLE III
SUBDIVISION OF THE GENUS *Lactobacillus*

| | *Growth at* | | |
|---|---|---|---|
| | *45 °C* | *15 °C* | |
| Thermobacterium | + | − | Homofermentative |
| Streptobacterium | ± | + | Homofermentative |
| Betabacterium | | | Heterofermentative |

The serological tests may be carried out using the methods of Sharpe (1955) or Mansi (1958). When testing for ammonia production from arginine, it is important to have a glucose concentration $\geq 2\%$ to prevent anomalous results.

(*i*) *Lactobacillus lactis*

Thermophilic starter, used in combination with other starters in production of Swiss type and Italian cheeses among others. Colonies normally rough, 1–3 mm in diameter and non-pigmented, being white to light grey. Metachromatic granules demonstratable with methylene blue.

Thermobacterium. Serology group E. Produces D(−) lactic acid. Ammonia not produced from arginine. Capable of fermenting salicin, sucrose and mannitol, but not amygdalin or cellobiose.

Requires some vitamins and amino acids as growth factors. Optimum growth temperature 40–43 °C.

(*ii*) *Lactobacillus bulgaricus*

Thermophilic starter, used in combination with other starters to produce yoghurt, Swiss-type and Italian cheeses (e.g. Grana) among others. Possesses metachromatic granules.

Thermobacterium. Serology group E. Produces D(−) lactic acid. Ammonia not produced from arginine. Very similar to *Lac. lactis*, but differs mainly in its inability to ferment such a wide range of sugars. Able to ferment lactose and cellobiose, but not amygdalin, maltose or mannitol.

Requires some vitamins and amino acids as growth factors. Optimum growth temperature is approximately 40 °C.

(*iii*) *Lactobacillus helveticus*

Thermophilic starter used in combination with other starters to produce Swiss-type and Italian cheeses among others. Colonies rough to rhizoid, 2–3 mm in diameter, and normally white to light grey. Possesses no metachromatic granules.

Thermobacterium. Serology group A. Produced DL lactic acid. Ammonia not produced from arginine. Able to ferment lactose, but not amygdalin, mannitol, salicin, sucrose or cellobiose.

Requires complex media. Optimum growth temperature 40–42 °C.

(*iv*) *Lactobacillus acidophilus*

Thermophilic starter for production of acidophilus milk and, with mesophilic starters, for production of Kefir. Colonies usually rough with no characteristic pigment. Possesses no metachromatic granules.

Thermobacterium. Strains appear serologically diverse and no group reactions have been demonstrated. Produces DL lactic acid. Ammonia not produced from arginine. Capable of fermenting amygdalin, cellobiose, lactose, salicin and sucrose but not mannitol. Produces the antimicrobial compounds acidophilin and acidolin, which are active against a wide range of Gram-positive and negative organisms.

Requires some vitamins and amino acids as growth factors. Optimum growth temperature 35–38 °C.

*(v) Lactobacillus casei*

Mesophilic starter. Used in production of Yakult. Also isolated from milk and milk products. Colonies may appear white to very light yellow. A number of sub-species have been recognised: *casei*, *alactosus*, *rhamnosus*, *tolerans* and *pseudoplantarum*.

Streptobacterium. Most strains belong to serology groups B and C. Produces L(+) lactic acid in excess of D(−) lactic acid. Ammonia not produced from arginine. Able to ferment amygdalin (except sub-sp. *tolerans*), cellobiose, mannitol (except sub-sp. *tolerans*), and salicin.

Unable to grow at 45 °C (except sub-spp. *rhamnosus* and *casei*).

*(vi) Lactobacillus plantarum*

Involved in the process of cheese ripening. Isolated from dairy products, silage, etc. Colonies white, but occasionally light or dark yellow. Some strains motile.

Streptobacterium. Serology group D. Produces DL lactic acid. Ammonia not produced from arginine. Able to ferment amygdalin, cellobiose, lactose mannitol, salicin and sucrose. Fermentation of melibiose and raffinose distinguishes this species from *Lac. casei*. Generally no growth at 45 °C.

Some amino acids and vitamins required for growth. Optimum growth temperature 30–35 °C.

*(vii) Lactobacillus curvatus*

Isolated from cow dung, dairy barn air, milk and silage. Colonies smaller but generally similar to *Lac. plantarum*. Some strains motile, but motility lost on subculture.

Streptobacterium. Produces DL lactic acid. Able to ferment cellobiose and salicin, but not amygdalin, mannitol or sucrose. Unlike *Lac. plantarum* is able to split aesculin. No growth at 45 °C.

Optimum growth temperature 30–37 °C.

*(viii) Lactobacillus fermentum*

Isolated from milk and fermented products. Colonies generally flat, circular or irregular to rough, often translucent.

Non-pigmented, but rare strains produce a rusty orange pigment.

Betabacterium. Serology group F. Ammonia produced from arginine. Can ferment lactose and sucrose, but not amygdalin, cellobiose, mannitol or salicin. Grows at 45 °C but not 15 °C.

Optimum growth temperature 41–42 °C for freshly isolated strains.

*(ix) Lactobacillus brevis*

Mesophilic starter, used in combination with other starters in the production of Kefir. Colonies generally rough and flat. May be translucent. Generally not pigmented, though some strains may produce orange to red colonies. Metrachromatic granules present.

Betabacterium. Serology group E. Ammonia produced from arginine. Unable to ferment amygdalin, cellobiose or salicin. Growth at 15 °C but not at 45 °C.

Complex nutritional requirements. Optimum growth temperature is approximately 30 °C.

*(x) Lactobacillus jugurti*

Used as a starter for yoghurt. No metachromatic granules present.

Thermobacterium. Serology group A. Produces DL lactic acid. Able to ferment lactose, but not amygdalin, cellobiose, mannitol, salicin or sucrose. Some strains may grow at 45 °C.

**The genus *Listeria***

*Listeria monocytogenes*

Pathogenic for man and cattle, e.g. causes abortion in cattle. Small coccoid rods (0·4–0·5 × 0·5–2·0 μm). Gram-positive, but may change as culture ages. Motile. Usually catalase-positive. On sheep liver extract, agar colonies appear smooth and milky by reflected light. Intermediary and rough colonies may also be observed, the latter arising from a filamentous form of growth. Some strains produce a yellow or red pigment after several months. Several types and sub-types may be distinguished by flagellar and somatic antigens, the latter exhibiting a partial antigenic relationship with other bacteria, but not with *Erysipelothrix* spp. Nitrates not reduced. Variable degrees of β-haemolysis depending on strain of organism and species of blood.

Requires a number of vitamins and growth factors.

Aerobic to micro-aerophilic. Minimum temperature 2·5 °C, killed at 58–59 °C in 10 min. Optimum growth temperature 37 °C.

## CORYNEFORM GROUP OF BACTERIA

This group includes the genera *Corynebacterium*, *Arthrobacter* with the related genera *Brevibacterium* and *Microbacterium*, *Cellulomonas* and *Kurthia*. Organisms belonging to most, if not all, of these genera have been isolated from milk. However, genera worthy of more detailed consideration here are: *Corynebacterium*, *Arthrobacter* and *Microbacterium*, although it should be noted that *Brevibacterium linens* is partly responsible for the development of characteristic flavour and aroma in surface ripened cheeses of the Limburger type (Pederson, 1971; Davis, 1976).

### General Characteristics of *Corynebacterium*, *Arthrobacter* and *Microbacterium*

Cells are straight to slightly curved rods (0·2–1·1 × 0·3–8·5 μm). Sometimes show club-shaped swellings. Arrangement of cells may be angular and palisade (picket fence). Usually non-motile. Gram-positive, although some species may stain Gram-negative. May contain irregularly stained segments or granules. Aerobic and facultatively anaerobic. Chemo-organotrophs.

### The Genus *Corynebacterium*

*Corynebacterium* spp. may be considered as belonging to one of three groups viz., (I) the human and animal pathogens, (II) the plant pathogens and (III) the non-pathogens. The *Corynebacterium* spp. principally associated with milk belong to the first of these groups, although since members of the non-pathogenic group have been isolated from soils, water and air, it is inevitable that they will contaminate milk. Analyses of milk flora include *Corynebacterium* spp. or 'coryneforms' among the organisms isolated (Gyllenberg *et al.*, 1963; Neill, 1974). Optimum growth temperature of the human and animal pathogenic species is 37 °C. Some species produce exotoxins.

Of these human and animal pathogens, *C. pyogenes* and *C. bovis* are worth mentioning.

#### (*i*) *Corynebacterium pyogenes*

Is thought to be the main causative organism of summer mastitis in cattle, a condition which, as the name of the organism suggests, is associated with the production of large amounts of pus. Colonies on blood agar are usually very small after 24 h, surrounded by large zones of clearing.

Produces a soluble haemolysin.
Growth is improved by the presence of blood or serum.

(*ii*) *Corynebacterium bovis*

This organism may exist as a commensal organism on cows' udders, but is thought possibly to cause mastitis (Cobb and Walley, 1962). Produces small white or cream coloured colonies on nutrient agar containing 1 % Tween 80.

Is not particularly fastidious, although some strains may have a requirement for nicotinic acid.

**The Genus *Arthrobacter***

Widely distributed in soil and, thus, like other organisms with this habitat, easily finds its way into milk. It is also very likely to be included under the general title of 'coryneforms' in data from milk flora analyses. In complex media, the cellular morphology of these organisms undergoes changes through the growth cycle. Older cells tend to be coccoid (0·4–0·9 μm in diameter), whereas younger cells are rod-shaped (0·4–0·8 × 1·0–6·0 μm). Gram-positive, but the rod forms may show only Gram-positive granules within Gram-negative cells. Optimum temperature is 20–30 °C, but most strains will grow at 10 °C. Strict aerobes.

Some species require growth factors.

**The Genus *Microbacterium***

Organisms belonging to this genus are found in milk, dairy products and on dairy equipment. They form part of the thermoduric flora (Seaman, 1963), and like *Arthrobacter* spp. and *Corynebacterium* spp., are usually labelled as 'coryneforms'. It is unlikely that they are particularly active as spoilage organisms. *M. lacticum* is probably the best known species in the dairy environment. Small rods (0·4–0·9 × 1·0–3·0 μm). Sometimes the arrangement of the cells may be similar to *Corynebacterium* spp. Gram-positive.

Are thermoduric, i.e. will survive heating at pasteurisation temperatures, and in fact will survive 72 °C for 15 min or more in skim-milk (Buchanan and Gibbons, 1974). Optimum temperature for growth is 30 °C.

## FAMILY PROPIONIBACTERIACEAE

The genera *Propionibacterium* and *Eubacterium* comprise this Family. Only the genus *Propionibacterium* is found associated with milk and dairy products.

**The genus *Propionibacterium***

Some species have been isolated from raw milk and other dairy products including cheese. *Propionibacterium freundenreichii* sub-species *shermanii* (*P. shermanii*) may be used along with *Lactobacillus* spp. as starter cultures in the production of Swiss or Emmenthaler cheese. Eye formation is caused by $CO_2$ produced by *P. shermanii* (Pederson, 1971). Cells are pleomorphic rods up to 6 μm in length. May occur in various arrangements. Gram-positive. Non-motile. Some strains are able to grow in the presence of 6·5 % NaCl or 20 % bile. Fermentation products include combinations of several organic acids (principally propionic and acetic acid) and $CO_2$ (Buchanan and Gibbons, 1974). Chemo-organotrophs. Anaerobic to aerotolerant. Grows best at 30–37 °C.

## FAMILY MYOBACTERIACEAE

This Family comprises only one genus, *Mycobacterium*.

Vary in morphology from coccoid to long filamentous forms depending on the strain, culture medium and environment. Size 0·2–0·6 μm × 1–10 μm. Not readily stainable by Gram's method, but usually considered Gram-positive. Acid- and alcohol-fast at some stage of growth. Non-motile. Species differ in catalase activity. All are aerobic, although from dispersed seeding in tubed agar medium, growth of some species occurs only in the depth of the medium. The most rapidly growing species require 2 to 3 days on a simple medium at 20–40 °C, and most pathogens require 2 to 6 weeks on complex media.

**The Genus *Mycobacterium***

Classical single test for identification is the demonstration of acid-fastness. Do not stain readily, but will accept hot acid dyes, e.g. carbol-fuchsin, and resist decolorisation with strong mineral acid and alcohol. Standard method of Ziehl Neelsen may be used. The degree of acid-fastness varies with strain, culture age and procedure used. This genus can be separated from the *Nocardia* by mycolic acid structure (Ratledge, 1977).

(*i*) *Mycobacterium tuberculosis*

Produces tuberculosis in man, but relatively non-pathogenic for goats, bovine animals or domestic fowls. Basic patterns of all forms of tuberculosis have been reviewed by Davies (1971). Slow grower, after 2 weeks on Lowenstein–Jensen (L–J) medium colonies appear rough becoming wrinkled and white. Attenuation of virulence may occur spontaneously upon subculture, but can be maintained by animal passage.

Niacin test positive. Catalase activity relatively weak, and lost after heating at 65 °C. Nitrate reduced. Growth inhibited by isoniazid, *p*-aminosalicylic acid and rifampin at 5 $\mu$g ml$^{-1}$ (for details see Mitruka, 1976). Strain to strain differences have been demonstrated by differing sensitivity to a number of phages.

Growth at 37 °C stimulated by incubation in air with 5–10 % added $CO_2$. No growth at 45 °C or 25 °C, some at 30–34 °C. Optimum growth temperature 37 °C.

(*ii*) *Mycobacterium bovis*

Produces tuberculosis in cattle, man and swine, but not pathogenic for most fowl. Infection in man is usually by ingestion of dairy products. Slow growers, after 3 weeks or more on L–J medium (preferably with the omission of glycerol), colonies appear scanty and white or buff coloured.

Niacin test negative. Growth inhibited by thiophene-2-carboxylic acid hydrazide (0·2 $\mu$g ml$^{-1}$), isoniazid and rifampicin at 0·5 $\mu$g ml$^{-1}$. Catalase-negative ($<$45 mm foam). Nitrate not reduced.

Micro-aerophilic when freshly isolated, but will adapt to aerobic conditions on repeated subculture. No growth at 25 °C or 45 °C, but able to grow at 37 °C and 40 °C.

## MISCELLANEOUS MICRO-ORGANISMS

### Bacteria

(*i*) *Campylobacter species*

Some species of the genus *Campylobacter* are pathogenic for man and animals, and may cause abortion or infertility in cattle and sheep, or disease in man, since some species can grow in the human intestinal tract (Buchanan and Gibbons, 1974). Several outbreaks of enteritis caused by *Campylobacter* spp. have been recorded resulting from the consumption of either raw or improperly pasteurised milk.

The genus *Campylobacter* belongs to the Family Spirillaceae. Cells are spirally curved rods (0·2–0·8 × 0·5–5 $\mu$m), and exhibit a corkscrew-like motion by means of a single flagellum at one or both poles. Gram-negative. Micro-aerophilic to anaerobic. Chemo-organotroph.

(*ii*) *Actinomyces bovis*

Causes lumpy jaw and actinomycosis in cattle. Pathogenicity for man not established. This organism belongs to the Family Actinomycetaceae. Gram-positive with irregular staining properties. Non-motile. Catalase-negative. Colonies (after 7–14 days in brain–heart infusion) are smooth and softer in consistency than those of *Act. israelii*, a human pathogen. Filaments with true branching may predominate, particularly in 18–48 h microcolonies. Diptheroid cells or branched rods are common. Serologically designated as group B with serotypes 1 and 2; the fluorescent-antibody technique is recommended to avoid cross reactions.

Homolactic fermentation. Capable of starch hydrolysis producing acid only. No nitrate reduction. Facultative anaerobe. Optimum growth temperature 35–37 °C.

(*iii*) *Nocardia rubropertincta*

Has been isolated from various soils, butter and air.

This organism belongs to the Family Nocardiaceae. Filamentation is abundant, usually after 14 h incubation, with the production, after 24 h, of microcysts which are more heat resistant than vegetative cells. Gram-positive. Partially acid-fast. Non-motile. Member of the Rhodochrous complex, there being considerable evidence that it should be raised to generic status (Bousfield and Goodfellow, 1976). Colonies on nutrient agar are elevated, rough, folded, dull with irregular margins and a coral red colour.

Acid produced from dextrin. Nitrate reduced by most strains. Strong DNAase activity. No gelatin liquefaction or peptonisation of litmus milk.

Capable of using a number of amino acids as sole carbon and energy sources. Obligate aerobe. Growth temperature range 10–40 °C. Optimum growth temperature 28–30 °C.

(*iv*) *Coxiella burnetii*

This is the causative organism of Q fever. These organisms are obligately intracellular parasites and grow preferentially in the vacuoles of host cells of both vertebrates and arthropods (especially ticks), the latter of which are vectors for the transmission of the disease (Q fever) among the former including cattle, sheep and goats. The disease has been reported in most countries of the world. Infected animals, which may be apparently healthy, may excrete the organism in their milk, and large numbers may also be released during parturition. Man may be affected by inhaling infected dust or, less commonly, by drinking contaminated raw milk, when a severe pulmonary infection may develop (Cruickshank *et al.*, 1973). However, Q fever is rarely fatal.

The genus *Coxiella* belongs to the Family Rickettsiaceae, cells are rod shaped or coccoid, and often pleomorphic (0·2–0·4 × 0·4–1·0 μm). They are sometimes seen as diplobacilli, and exhibit a filterable phase in their developmental cycle (Buchanan and Gibbons, 1974).

The genus is characterised by its great resistance to physical and chemical agents in an extracellular environment. *Coxiella burnetii* is remarkably resistant to desiccation, and will survive for long periods in the tissues or faeces of infected ticks and in wool, etc. The organism remains viable for several days in water or milk. It is fairly resistant to most of the commonly used antiseptics, and will survive exposure to 1 % phenol for 1 h. Although *C. burnetii* may not be killed by fairly severe heat treatments, heating at 62·75 °C for 30 min or 71·75 °C for 15 s (conventional pasteurisation treatments) is sufficient to render raw milk free from this organism (Enright *et al.*, 1957).

**Yeasts**

The following are a number of yeasts which have been associated with milk. The morphological descriptions have been taken from Lodder (1971).

*(i) Debaryomyces hansenii*

Has been isolated from cheese among other foodstuffs. In malt extract broth (2 days at 25 °C), cells appear spherical to short oval and are arranged singly, in pairs or in short chains, and propagate by multipolar budding. On slide cultures pseudomycelium is absent or, if present, is very primitive, only on rare occasions is it well developed. Ascospores are spherical with one or two per ascus. It is unable to ferment lactose, but may be able to assimilate it.

*(ii) Kluyveromyces fragilis*

Has been isolated from yoghurt, and is associated with the manufacture of Kefir and Kumiss. In malt extract broth (3 days at 28 °C), cells appear subglobose, ellipsoidal to cylindrical and occur as single cells, in pairs or in short chains. Varying degrees of pseudomycelium are observed. The ascospores are stoutly kidney-shaped or oblong with obtuse ends, one to four being produced per ascus. Lactose is rapidly fermented with the production of much $CO_2$ and little alcohol.

*(iii) Kluyveromyces lactis*

Associated with yoghurt, and has been isolated from cheese, buttermilk and cream. In malt extract broth (3 days at 28 °C), cells appear spherical, ellipsoidal or occasionally cylindrical. They are arranged singly, in pairs or occasionally in small clusters. Amoeboid cells may be formed. The ascospores are spherical to prolate ellipsoidal, and one to four may be produced per ascus. Is able to ferment lactose and galactose.

(*iv*) *Saccharomyces cerevisiae*

Used in the brewing (called a 'top yeast') and baking industries. Has been isolated from Stracchino cheese. In malt extract broth (3 days at 28 °C), cells may appear spherical, subglobose, ovoid, ellipsoidal or cylindrical to elongate and occur singly, in pairs and occasionally in short chains or small clusters. Ascospores are round, smooth, and usually one to four are produced per ascus. In general, ascus formation is not immediately preceded by conjugation or the formation of cells with evaginations. Glucose but not lactose is fermented.

(*v*) *Candida lipolytica var. lipolytica*

Has been isolated from butter and margarine. When grown in glucose yeast extract peptone water (3 days at 25 °C), cells appear short ovoid to elongate. On slide cultures, a well developed pseudomycelium and some true mycelium are produced. Capable of hydrolysing fats but unable to ferment sugars, though erythritol can be assimilated.

(*vi*) *Candida kefir*

Associated with buttermilk, cheese and Kefir. When grown in glucose yeast extract peptone water (3 days at 25 °C), cells appear short ovoid to long ovoid. Pseudomycelium is abundant in slide cultures. Lactose and glucose fermented, but characteristically D-xylose is not assimilated.

(*vii*) *Torulopsis lactis-condensi*

Isolated from sweetened condensed milk. When grown in glucose yeast extract peptone water (3 days at 25 °C), cells appear ovoid. Capable of fermenting glucose but not lactose, mannitol and glucitol are not assimilated.

**Moulds**

More detailed descriptions of the morphology of these organisms can be found in Smith (1969).

(*i*) *Geotrichum candidum*

Commonly found on dairy products. On malt extract agar, colonies are white in colour and appear yeast-like and butyrous, particularly when older. Forms a true mycelium which breaks up almost completely into arthrospores which are cylindrical in shape with rounded ends.

(*ii*) *Scopulariopsis brevicaulis*

Grows best on substrates high in protein, e.g. cheese, and is a serious source of infection in mould-ripened cheese, where it gives rise to an ammoniacal taste and odour. Colonies usually appear thin and smooth velvety at first (often furrowed), greyish-white in colour, then becoming yellowish-brown and overgrown with loosely floccose to funiculose hyphae. The conidia are most characteristic, being truncated spheres with a thickened basal ring around the truncation.

(*iii*) *Sporendonema sebi*

Forms discrete colonies (cf. buttons) in sweetened condensed milk. Forms endogenous spores, i.e. formed within the conidiophore, which 'round up' and are eventually released when mature.

(*iv*) *Penicillium roqueforti*

Used in the manufacture of Stilton, Roquefort, Gorgonzola and similar blue-veined cheeses. Belongs to the Asymmetrica group, i.e. more than one branch in the conidiophore, and this branching being asymmetrical. All strains grow well on malt agar. Usually produce blue-green spreading colonies changing to a darker green. The colonies have a smooth velvety appearance with an irregular margin of radiating lines of conidiophores, an effect that has been compared to a 'spiders web' and termed arachnoid. The conidiophores are rough, and the conidia are globose, smooth and borne in loose columns or tangled chains.

(*v*) *Penicillium casei*

Found associated with some Swiss cheeses. Member of the Asymmetrica group. Similar to *P. roqueforti*, the differences being slower growth rate, lack of arachnoid margin and the reverse of the colonies being yellow-brown instead of green.

(*vi*) *Penicillium camemberti and Penicillium caseicola*

Important in the production of Camembert, Brie and similar cheeses. Members of Asymmetrica group. Colonies of *P. caseicola* remain white whereas those of *P. camemberti* gradually become pale greyish-green from the centre outwards. Penicilli are irregular with rami and metulae at the same level with few branches at each stage. Conidiophores are slightly rough. Conidia become subglobose and are borne in tangled chains.

*(vii) Rhizopus stolonifer*

Frequently found on stale bread and many other foodstuffs. Grows rapidly on many substrates except, because of its inability to utilise nitrate, Czapek Dox agar. Sporangiophores arise from nodes at which thick tufts of rhizoids develop. Sporangia are globose and white at first, becoming black as spores ripen.

**Viruses**

*(i) Cow-Pox Virus: Genus Poxvirus Sub-genus A*

Cows are not usually seriously affected, but the virus produces, on their teats, vesicles or pustules which may break during milking leaving raw tender areas. Crusting follows, with the scabs falling off in 10 days leaving an unscarred surface. Can be transmitted to the hands of milkers producing lesions on the back of the hands, forearms and face.

In ultra-thin sections of infected cells, the virus (virion) appears oval with a multilayered covering, although it has no true envelope. Contains double-stranded DNA. Similar to vaccina virus, but has slight cultural and serological differences.

*(ii) Poliomyelitis: Genus Enterovirus*

Usually the patient has no apparent symptoms, but may experience fever, headache and vomiting. In a small proportion of cases, the virus becomes localised in the central nervous system leading to paralysis. May be spread by inhalation, or by ingestion of foodstuffs contaminated (e.g. by flies) with faecal material from a patient or a carrier of the disease.

Virus has an icosohedral shape, 27 nm in diameter, enclosing a single-stranded RNA core. No envelope. It is characterised by a strong affinity for nervous tissue and a narrow host range, primates only being readily susceptible. Three serotypes found: Type 1, Brunhilde and Mahoney strains; Type 2, includes Lansing and MEF 1 strains and Type 3, Leon and Sankett strains. Readily killed by moist heat at 50–55 °C, but milk may exert a protective effect, e.g. may survive 60 °C. Pasteurisation at approximately 72 °C (HTST) is preferred to give a satisfactory safety margin.

*(iii) Central European Tick-Borne Fever: Genus Flavovirus*

Produces a bi-phasic illness with an influenza-like onset followed by a period of apyrexia for 4–10 days, and culminating in meningitis or minings-encephalitis (Cruickshank, 1965). Transient paralysis is not infrequent.

Occurs widely in Central Europe and in the USSR, the viruses causing the disease in these respective areas are closely related, but antigenically distinct. Vectors are the ticks, *Ixodes ricinus* and *Ixodes persulcatus*. About 23 % of human cases have been attributed to drinking unboiled goat's milk (Cruickshank, 1965). Virus is spherical, 20–50 nm in diameter, and encloses a single-stranded RNA core. Envelope present. Killed by heating at 60 °C for 10 min.

(*iv*) *Hepatitis*

Hepatitis in man can be caused by a variety of viruses, e.g. rubella and adenovirus, as well as non-viral organisms, e.g. *Coxiella burnetti* and toxoplasma but, particularly in temperate regions, the majority of cases are caused by hepatitis virus A (infective hepatitis) or B (serum hepatitis). Hepatitis B virus is distinct from hepatitis A virus, and is found in the blood of cases and carriers, being spread by tissue penetration with inadequate asepsis, e.g. during blood transfusion, and in kidney dialysis units. The following description is restricted to infective hepatitis which is essentially a disease of children, and can be spread by the ingestion of food contaminated with faeces from a case or carrier of the disease. It has been known to cause epidemics in schools, institutions and military camps, particularly where sanitation is poor. The symptoms of the disease are fever, malaise and jaundice, but in only a few cases does multilobular cirrhosis of the liver occur (Cruickshank, 1965).

To the authors' knowledge the size is unknown, but is thought to be 12–18 nm in diameter and to enclose RNA. It can survive 56 °C for 30 min (Cruickshank *et al.*, 1973).

(*v*) *Bacteriophages of Lactic Acid Bacteria*

Bacteriophages (phages) are one of the causes of the slow growth of starter cultures. Although some phages are species-specific, there is a large group which are non-species-specific (Chopin *et al.*, 1976).

Keogh and Shimmin (1974) isolated a number of phages from *Str. lactis*, *Str. lactis* sub-sp. *diacetylactis* and *Str. cremoris* and found that all had either a prolate polyhedral or octahedral head, with most having non-contractile tails, type III tail plates and belonged to group IV in tne Tikhonenko (1970) classification. This is in substantial agreement with Law and Sharpe (1978), who also stated that all lactic acid phages contained a DNA core. Huggins and Sandine (1977) found that out of 38 temperate phages from lactic streptococci, all had isometric heads and non-contractile tails; some had collars and structurally distinctive base plates.

To prevent slow growth of a starter the culture may be grown, prior to inoculation into the batch of milk, in a medium low in $Ca^{2+}$ ions which are required for the adsorption of phage to the host cell. The use and frequent rotation of mixed starters, consisting of phage-unrelated strains, is also employed to prevent any serious build up of phage particles.

## REFERENCES

ANATSKAYA, A. G. and EFIMOVA, V. A. (1978) *Brief Communications of XXth International Dairy Congress*, **E**, 739.

BABEL, F. J. (1977) *J. Dairy Sci.*, **60**, 815. Cited in: Sharpe, M. E. (1979), **32**, 9.

BAIRD-PARKER, A. C. (1963) *J. Gen. Microbiol.*, **30**, 409. Cited in: Gibbs, B. M. and Skinner, F. A. (1966) *Identification Methods for Microbiologists, Part A*, Academic Press Ltd, London.

BERGÈRE, J.-L. (1978) *Brief Communications of the XXth International Dairy Congress*, **E**, 766.

BERGÈRE, J.-L. and HERMIER, J. (1970) *J. Appl. Bacteriol.*, **33**, 167.

BOUSFIELD, I. J. and GOODFELLOW, M. (1976). The 'rhodochrous' complex and its relationships with allied taxa. In: *The Biology of the Nocardiae*, Goodfellow, M., Brownell, G. H. and Serrano, J. A. (Eds.) Academic Press Ltd, London.

British Standards Institution (1968) BS 4285 *Methods of Microbiological Examination for Dairy Purposes*, BSI, London.

BUCHANAN, R. E. and GIBBONS, N. E. (1974) *Bergey's Manual of Determinative Bacteriology*, 8th edn., Williams & Wilkins Co., Baltimore.

BURTON, H. (1977) *J. Soc. Dairy Technol.*, **30**, 135.

CANDY, M. R. and NICHOLS, A. A. (1956) *J. Dairy Research*, **23**, 329. Cited in: Hersom, A. C. and Hulland, E. D. (1969) *Canned Foods*, 6th edn., J. & A. Churchill Ltd, London.

CHOPIN, M.-C., CHOPIN, A. and ROUX, C. (1976) *Appl. and Environmental Microbiol.*, **32**, 741. Cited in: Skinner, F. A. and Quesnel, L. Q. (1978) *Streptococci*, Academic Press Ltd, London.

CLARKE, P. H. and RICHMOND, M. H. (1975) *Genetics and Biochemistry of* Pseudomonas, John Wiley & Sons, London.

COBB, R. W. and WALLEY, J. K. (1962) *Veterinary Record*, **74**, 101. Cited in: Buchanan, R. E. and Gibbons, N. E. (1974) *Bergey's Manual of Determinative Bacteriology*, 8th edn., Williams & Wilkins Co., Baltimore.

CROSSLEY, E. L. (1946) *J. Dairy Research*, **14**, 233. Cited in: Foster, E. M., Nelson, F. E., Speck, M. L., Doetsch, R. N. and Olson, J. C. (1958) *Dairy Microbiology*, Macmillan & Co. Ltd, London.

CRUICKSHANK, R. (1965) *Medical Microbiology*, 11th edn., E. & S. Livingstone Ltd, Edinburgh and London.

CRUICKSHANK, R., DUGUID, J. P., MARMION, B. P. and SWAIN, R. H. A. (1973) *Medical Microbiology*, 12th edn., Churchill Livingstone Ltd, London.

DAVIES, P. D. B. (1971) *Brit. J. Hospital Medicine*, **5**, 749. Cited in: Ratledge, C. (Ed.) (1977) *The Mycobacteria* (*Patterns of Progress Series*), Meadowfield Press Ltd, England.

DAVIS, J. G. (1965) *Cheese* (Vol. I) J & A Churchill Ltd, London.

DAVIS, J. G. (1976) *Cheese* (Vol. III) Churchill Livingstone Ltd, London.

DEMPSTER, J. F. (1968) *J. Appl. Bacteriol.*, **31**, 290.

Department of Health and Social Security (1969) *The Bacteriological Examination of Water Supplies*, Report No. 71, HMSO, London.

DERBY, H. A. and HAMMER, B. W. (1931) *Iowa Agricultural Experimental Station Research Bull.*, **145**, 387.

DEWBERRY, E. B. (1959) *Food Poisoning*, 4th edn., Leonard Hill (Books) Ltd, London.

EDWARDS, S. J. (1933) *J. Comparative Pathol. and Therapeutics*, **46**, 211. Cited in: Harrigan, W. F. and McCance, M. E. (1966) *Laboratory Methods in Microbiology*, Academic Press Ltd, London.

ELEK, S. D. and LEVY, E. (1950) *J. Pathologic. Bacteriol.*, **62**, 541. Cited in: Buchanan, R. E. and Gibbons, N. E. (1974) *Bergey's Manual of Determinative Bacteriology*, 8th edn., Williams & Wilkins Co., Baltimore.

ENRIGHT, J. B., SADLER, W. W. and THOMAS, R. C. (1957) *Am. J. Public Health*, **47**, 695. Cited in: Cruickshank, R. (1965) *Medical Microbiology*, 11th edn., E. & S. Livingstone Ltd, Edinburgh and London.

FOSTER, E. M., NELSON, F. E., SPECK, M. L., DOETSCH, R. N. and OLSON, J. C. (1958) *Dairy Microbiology*, Macmillan & Co. Ltd, London.

GYLLENBERG, H. G., EKLUND, E., CARLBERG, G., ANTILA, M. and VARTIOVAARA, U. (1963) *Acta Agriculturae Scandinavica*, **13**, 177.

HAMON, Y. (1956) *Annales de l'Institut Pasteur*, **91**, 489. Cited in: Clarke, P. H. and Richmond, M. H. (1975) *Genetics and Biochemistry of* Pseudomonas, John Wiley & Sons, London.

HAMON, Y., VERON, M. and PERON, Y. (1961) *Annales de l'Institut Pasteur*, **101**, 738. Cited in: Clarke, P. H. and Richmond, M. H. (1975) *Genetics and Biochemistry of* Pseudomonas, John Wiley & Sons, London.

HARRIGAN, W. F. and MCCANCE, M. E. (1966) *Laboratory Methods in Microbiology*, Academic Press Ltd, London.

HENDRIE, M. S. and SHEWAN, J. M. (1966) The identification of certain *Pseudomonas* species. In: *Identification Methods for Microbiologists, Part A*, Gibbs, B. M. and Skinner, F. A. (Eds.), Academic Press Inc., London.

HERSOM, A. C. and HULLAND, E. D. (1969) *Canned Foods*, 6th edn., J & A Churchill Ltd, London.

HOBBS, B. C. and GILBERT, R. J. (1978) *Food Poisoning and Food Hygiene*, 4th edn., Edward Arnold, London.

HUGGINS, A. R. and SANDINE, W. E. (1977) *Appl. Environmental Microbiol.*, **33**, 184.

HUGHES, D. (1979) *J. Appl. Bacteriol.*, **46**, 125.

JAYNE-WILLIAMS, D. J. and FRANKLIN, J. G. (1960) *Dairy Sci. Abstr.*, **22**, 215.

JEFFRIES, L. (1969) *Internat. J. Systematic Bacteriol.*, **19**, 183. Cited in: Buchanan, R. E. and Gibbons, N. E. (1974) *Bergey's Manual of Determinative Bacteriology*, 8th edn., Williams & Wilkins Co., Baltimore.

JONES, D. (1978) Composition of streptococci. In: *Streptococci*, Skinner, F. A. and Quesnel, L. B. (Eds.), Academic Press Ltd, London.

JONES, D., DEIBEL, R. H. and NIVEN, C. F., JR. (1963) *J. Bacteriol.*, **85**, 62. Cited in: Buchanan, R. E. and Gibbons, N. E. (1974) *Bergey's Manual of Determinative Bacteriology*, 8th edn., Williams & Wilkins Co., Baltimore.

JUFFS, H. S. (1973) *J. Appl. Bacteriol.*, **36**, 585.
KEOGH, B. P. and SHIMMIN, P. D. (1974) *Appl. Microbiol.*, **27**, 411.
KING, A. and PHILLIPS, I. (1978) *J. Medical Microbiol.*, **11**, 165.
KING, E. O., WARD, M. K. and RANEY, D. E. (1954) *J. Laboratory Clinical Medicine*, **44**, 301.
KORHONEN, H., ALI-YRKKÖ, S. and HALKARAINEN, H. (1978) *Brief Communications of the XXth International Dairy Congress*, **E**, 764.
LANCEFIELD, R. C. (1933) *J. Experimental Medicine*, **57**, 571. Cited in: Gibbs, B. M. and Skinner, F. A. (1966) *Identification Methods for Microbiologists, Part A*, Academic Press Ltd, London.
LAUTROP, H. Unpublished information. Cited in: Buchanan, R. E. and Gibbons, N. E. (1974) *Bergey's Manual of Determinative Bacteriology*, 8th edn., Williams & Wilkins Co., Baltimore.
LAW, B. A. and SHARPE, M. E. (1978) Dairy Industry Streptococci. In: *Streptococci*, Skinner, F. A. and Quesnel, L. B. (Eds.), Academic Press Ltd, London.
LEVIN, R. E. (1968) *Appl. Microbiol.*, **16**, 1734.
LIGHTBODY, L. G. and PETERSEN, H. P. (1962) *Queensland J. Agric. Sci.*, **19**, 373. Cited in: Thomas, S. B. and Druce, R. G. (1971) *Dairy Industries*, **36**, 75.
LODDER, J. (1971) *The Yeasts*, North-Holland Publishing Company, London.
LONG, H. F. and HAMMER, B. W. (1941) *Research Bull. Iowa Agricultural Experimental Station*, **285**, 176.
LYSENKO, O. (1961) *J. Gen. Microbiol.*, **25**, 379.
MALMO, J., ROBINSON, B. and MORRIS, R. S. (1972) *Australian Veterinary J.*, **48**, 137. Cited in: Amemiya, J., Takase, K. and Sato, H. (1978) *The Bull. of the Faculty of Agriculture, Kagoshima University*, No. 28.
MANSI, W. (1958) *Nature, London*, **181**, 1289. Cited in: Gibbs, B. M. and Skinner, F. A. (1966) *Identification Methods for Microbiologists, Part A*, Academic Press Ltd, London.
MATTEUZZI, D., TROVATELLI, L. D., BIAVATI, B. and ZANI, G. (1977) *J. Appl. Bacteriol.*, **43**, 375.
MIKOLAJCIK, E. M. (1978) *Am. Dairy Rev.*, **40**, 34A.
MITCHELL, R. G. and BAIRD-PARKER, A. C. (1967) *J. Appl. Bacteriol.*, **30**, 251. Cited in: Buchanan, R. E. and Gibbons, N. E. (1974) *Bergey's Manual of Determinative Bacteriology*, 8th edn., Williams & Wilkins Co., Baltimore.
MITRUKA, B. M. (1976) *Methods of Detection and Identification of Bacteria*, CRC Press Inc., Cleveland.
MORRISON, H. B. and HAMMER, B. W. (1941) *J. Dairy Sci.*, **24**, 9.
NEILL, S. D. (1974) *A study of the microflora of raw milk stored at low temperatures*, Ph.D. thesis, Queen's University, Belfast.
ORLA JENSEN, S. (1919) *The Lactic Acid Bacteria*, Andr. Fred Host & Son, Copenhagen. Cited in: Gibbs, B. M. and Skinner, F. A. (1966) *Identification Methods for Microbiologists, Part A*, Academic Press Ltd, London.
ORLA JENSEN, S. (1943) *The Lactic Acid Bacteria*, Einar Munksgaards, Copenhagen. Cited in: Gibbs, B. M. and Skinner, F. A. (1966) *Identification Methods for Microbiologists, Part A*, Academic Press Ltd, London.
PATTERSON, A. C. (1965) *J. Gen. Microbiol.*, **39**, 295. Cited in: Clarke, P. H. and Richmond, M. H. (1975) *Genetics and Biochemistry of* Pseudomonas, John Wiley & Sons, London.

PECKNOLD, P. C. and GROGAN, R. G. (1973) *Internat. J. Systematic Bacteriol.*, **23**, 111. Cited in: Palleroni, N. J. (1978) *The* Pseudomonas *Group* (*Patterns of Progress Series*), Meadowfield Press Ltd, England.

PEDERSON, C. S. (1971) *Microbiology of Food Fermentation*, AVI Publishing Co Inc., Connecticut.

RATLEDGE, C. (Ed.) (1977) *The Mycobacteria* in: Cook, J. G. (Ed.) (1977) *Patterns of Progress Series*, Meadowfield Press Ltd, England.

ROGUINSKY, M. (1969) *Annales de l'Institut Pasteur*, **117**, 529. Cited in: Skinner, F. A. and Quesnel, L. B. (1978) *Streptococci*, Academic Press Ltd, London.

ROGUINSKY, M. (1971) *Annales de l'Institut Pasteur*, **120**, 154. Cited in: Skinner, F. A. and Quesnel, L. B. (1978) *Streptococci*, Academic Press Ltd, London.

SCHLEIFER, K. H. and KLOOS, W. E. (1975) *Internat. J. Systematic Bacteriol.*, **25**, 50.

SEAMAN, A. (1963) *Bacteriology for Dairy Students*, Cleaver-Hume Press Ltd, London.

SHARPE, M. E. (1955) *J. Gen. Microbiol.*, **12**, 107. Cited in: Gibbs, B. M. and Skinner, F. A. (1966) *Identification Methods for Microbiologists, Part A*, Academic Press Ltd, London.

SHERMAN, J. M. (1937) *Bacteriologic. Rev.*, **1**, 3.

SKADHAUGE, K. (1950) *Studies of Enterococci*, Einar Munksgaards, Copenhagen. Cited in: Buchanan, R. E. and Gibbons, N. E. (1974) *Bergey's Manual of Determinative Bacteriology*, 8th edn., Williams & Wilkins Co., Baltimore.

SMITH, G. (1969) *An Introduction to Industrial Mycology*, Academic Press Ltd, London.

STERN, N. J. and PIERSON, M. D. (1979) *J. Fd Sci.*, **44**, 1736.

STEWART, D. B. (1975) *J. Soc. Dairy Technol.*, **28**, 80.

STEWART, D. B. (1978) *Brief Communications of the XXth International Dairy Congress*, **E**, 91.

THOMAS, S. B. (1958) *Dairy Sci. Abstr.*, **20**, 357, 449.

THOMAS, S. B. and THOMAS, B. F. (1973) *Dairy Industries*, **38**, 61.

TIKHONENKO, A. S. (1970) *Ultrastructure of Bacterial Viruses*, Plenum Press, New York. Cited in: Keogh, B. P. and Shimmin, P. D. (1974) *Appl. Microbiol.*, **27**, 411.

WITTER, L. D. (1961) *J. Dairy Sci.*, **44**, 983.

# 3

# Control and Destruction of Micro-Organisms

ROBERT R. ZALL
*Department of Food Science, Cornell University, Ithaca, New York, USA*

Before considering some of the different methods available to control or destroy micro-organisms in milk and milk products, it is important to understand how micro-organisms find their way into dairy products. It is also useful to remember that milk constituents provide an almost ideal substrate in which micro-organisms can grow and multiply.

The bacterial flora in milk can vary considerably in numbers and species depending on how the milk is soiled. Thus, milk contains few bacteria in the cow's udder, but is later subject to contamination by man and his habits. He contaminates milk by his animal husbandry methods, and by the practices he uses in collecting milk from the cows, but both the numbers and kinds of contaminants to be found in milk also depend on the health of the animals. Of special interest to us at this time are the psychrotropic bacteria which grow in refrigerated milk, and much of our milk is now held longer and colder since the advent of on-farm bulk milk handling–holding systems. Milk and milk products are subject to additional contamination with different kinds of micro-organisms at milk collection stations, and because milk tends to be older with current milk handling methods, new problems are coming to light.

Some finished goods contain sugar, spices, flavours, salt, stabilisers, etc., and each condiment in itself may contribute additional varieties of micro-organisms. Initially, therefore, we can look at milk as a product with contaminating micro-organisms in two broad categories; those organisms that are found in raw milk, and those bacteria to be found in pasteurised milk and milk products.

## RAW MILK

The introduction of bulk milk (farms equipped with refrigerated tanks where the product is stored for two or more milkings) means that milk does not have to go to a factory daily, and the milk can be picked up by an over-the-road truck for transport to the milk factory at times more convenient to the processor. Because milk is held longer and colder in its raw state, a situation is created where cold-tolerant organisms can grow. This group of organisms is known as the psychrotrophs, and is capable of appeciable growth at 2–7 °C, irrespective of the optimum growing temperature. The most widely encountered genera in this group are: *Pseudomonas*, *Flavobacterium*, *Alcaligenes* and *Achromobacter*. We also find *Streptococcus*, yeasts, moulds and *Bacillus coagulans*.

Elsewhere, contaminants get into milk via dairy utensils and milk contact surfaces. These sources contribute lactic streptococci, coliform organisms, and the Gram-negative psychrotrophs, while in addition, milking equipment provides excellent locations in which thermoduric bacteria can thrive and contaminate milk. Surely the milk handlers themselves contribute additional bacteria to the 'pot-pourri' of micro-organisms that can and do get into our milk supply, and once these different kinds of bacteria get into milk, then their numbers increase over time.

## CONTROL

It is neither easy nor efficient to control or destroy micro-organisms after they get into milk, as opposed to preventing contamination in the first place. Problems arising from the presence of psychrotrophs, antibiotics and adventitious chemicals are perhaps the most frequently encountered in the dairy industry, and among these, the advent of psychrotrophs is probably the most serious, because once they grow in refrigerated milk, the milk can develop serious off-flavours.

Much has been written about handling milk properly, including the use of good sanitation procedures in processing; prompt and sufficient cooling have also been emphasised. The US Public Health Service (1965) has promulgated explicit regulations for the handling of milk including cooling, and it sets limits for bacterial numbers in milk as do regulatory agencies in other countries. In spite of these edicts, we continue to get reports that pasteurised milks are not holding-up long enough during distribution, and are deteriorating unreasonably. According to a 1979 year-round survey on pasteurised milk quality, Bandler (1979) reported that only 20 % of New York State's sell-by date samples have bacterial counts less than 20 000 ml$^{-1}$. Organoleptically, 92·5 % of them either lack freshness, have fruity (psychrotrophic) flavours, or are rancid. Even processors employing ultra-high temperature (UHT) heat treatments find sterilised milk having problems, such as developing bitterness, clearing, or coagulation (Hsu, 1970).

## PASTEURISED PRODUCT

Recontamination of milk (post-pasteurisation) may be one of the factors that speeds up milk deterioration in finished goods, but the presence of psychrotrophs in cold raw milk (preprocessing) could be the critical factor in undermining the keeping quality of pasteurised milk and other dairy products. Speck and Adams (1976) pointed out that a psychrotroph population in milk under 10 000 ml$^{-1}$ can still produce about 10 or more units of heat-stable proteases per ml, and there is little doubt that heat-stable lipases or proteases produced by psychrotrophs can remain active after pasteurisation of the milk. Thus, it is common knowledge that lipases of milk origin are heat labile, and by simply heating milk at 72 °C for 5 s inactivates over 85 % of the enzyme (Nilsson and Willart, 1960). On the other hand, bacterial lipases are very heat resistant. Hedlund (1976) reported that when a pseudomonad lipase was added to milk, only 45 % of the enzyme was inactivated by a heat treatment of 105 °C for 10 min. Such lipases are thought to cause rancidity in milk, and their presence in milk can be extremely objectionable. One measurement of the degree of rancidity in milk is its 'acid degree value' (ADV), which measures the free fatty acids liberated when milk fats are hydrolysed. Normal milk has ADVs ranging from 0·4 to 0·8, and a rancid taste begins to appear as a flavour defect to sensitive people when milk registers an ADV of 1·0 to 1·4. When the value exceeds 1·5, milk is rancid in taste to most people.

Besides micro-organisms in milk, milk contains enzymes which are present either naturally, being produced in the animal's mammary gland, or they can be formed as extracellular products by growing micro-organisms, or become bothersome residues when micro-organisms are lysed. In countries where much of the milk is converted into cheese products, enzymes from microbial contaminants cause serious problems, for they not only interfere with delicate cheese flavours, but also cause economic losses to the cheese industry by decreasing product yields. According to the complaints of individual cheesemakers, different nationally identified cheeses, like British Cheddar, Italian Romano and others, are affected by lipases and proteases in different ways, and it is quite accurate to think that the biological and enzymatic properties of micro-organisms in milk need to be regulated to act in a more or less uniform way by treating milk with heat and chemicals.

## FACTORS AFFECTING GROWTH

Growth requirements for bacteria vary from species to species. Thus, some bacteria can synthesise vitamins, while others must have them available in their substrate, and it is also true that micro-organisms require other essential metabolites varying from simple ions, i.e. $Cu^{2+}$ or $Fe^{3+}$, to a full range of complex organic substances. Micro-organisms need this food for growth and reproduction, and to most bacteria and fungi, a food supply usually means an energy supply; one exception being photosynthetic bacteria. To deny micro-organisms nutrients means that they are unable to obtain energy materials from which they make up cellular material, and the essential minerals needed for them to function. With this in mind, food technologists ought to consider the primary factors that affect microbial  growth, which are: (1) moisture, (2) nutrients, (3) oxygen, or lack of it, (4) temperature and (5) pH. By regulating or manipulating these environmental factors in a food system, the dairy scientist can control microbial growth.

Microbiologists do think in terms of micro-environment which, as a concept, obtained favourable acceptance in the mid-sixties. A foodstuff such as cheese, for example, may not be uniform in composition, so that different micro-organisms grow in different parts of a cheese substrate, and may be stressed by different environmental factors. A case in point can be salt. Thus, salt in water affects microbial growth and, in fact, bacteriologists classify bacteria by their ability/inability to grow in different substrates with varied percentages of salt; dissolved salt alters osmotic pressure, and this factor will selectively control microbial growths. Adding 1–2% salt to cheese does not mean that a salt will be distributed uniformly and, in fact, the water phase in the cheese might contain a concentration of salt 10 times that added to the mass. The fact that different species of organism possess different tolerances to salt, means that this mechanism can be a useful tool to control, manipulate or destroy micro-organisms.

This discussion leads us to the concept of water activity. Microorganisms need water to grow, and this water requirement can be expressed in terms of water activity ($A_w$). The term expresses the degree of water availability, and is most frequently used in describing foodstuffs; specifically, it is a measurement of the vapour pressure of a solution divided by the vapour pressure of the solvent. Water has an $A_w$ of 1·0, whereas a 1·0 molar solution of an ideal solute has an $A_w$ of 0·9823, and this reduction of $A_w$ results from the solute dissolving in water; the freezing point of the water is also lowered. Fresh foods which are subject to biological spoilage usually show $A_w$ values about 0·95, but the $A_w$ can be lowered by adding salt or sugar, which then makes the food less vulnerable to spoilage. Table I illustrates some water activity values tolerated by food spoilage micro-organisms, and some other micro-organisms can grow in quite high concentrations of salts and sugars. A most notable group is the halophilic

TABLE I
WATER ACTIVITY VALUES FOR SELECT GROUPS OF MICRO-ORGANISMS

| *Micro-organisms* | *Common $A_w$ values* |
|---|---|
| Normal bacteria | 0·91 |
| Normal yeasts | 0·88 |
| Normal moulds | 0·80 |
| Xerophilic fungi | 0·65 |
| Osmophilic yeasts | 0·60 |

bacteria, capable of growth in solutions with $A_w$ values approximating 0·7, and these organisms are frequently isolated in fishery wastes and in ocean water. Aside from salt tolerant species, we encounter osmophilic yeasts which bother liquid sugar products, and these have to be controlled in our dessert and ice cream factories where we use liquid sugars.

It is, therefore, logical to think in terms of limiting microbial growth by controlling moisture, but as Table I shows, moulds and yeasts survive better with less water than bacteria. These data also imply that microbial growth can be lessened by creating situations where the availability of moisture is decreased, and milk powder is one example of a method of preserving milk solids by reducing moisture; the moisture content in a Grade A milk powder might vary from 3–4%.

## DESTRUCTION BY HEAT

It is well known that bacteria survive in temperatures between −250 °C to 160 °C, and that micro-organisms appear to have minimum, maximum and optimum temperatures at which they grow. In addition, researchers of many years past learned that micro-organisms have a thermal death point; that is to say that organisms vary in the time–temperature relationship at which they can be destroyed by heat. A thermal death point is the length of time needed to kill organisms at a given temperature in a given material (this is a critical point for concern to firms canning foodstuffs).

Intuitively, one must recognise the fact that, in the destruction of micro-organisms by heat, such factors as the kind of substrate, age of micro-organism, pH, etc., all have to be considered. Nevertheless, organisms do die in an orderly way, and it is possible to predict their destruction by plotting a series of survivor curves. The order of death is pretty much the same for all unicellular organisms, and is considered to be logarithmic in nature. Some explain the fact that a logarithmic kill of bacteria by heat occurs because heat destroys the genes governing reproduction, and this destruction appears to take place as a result of protein denaturation. Almost fifty years ago, Otto Rahn (1945) published a monograph dealing with injury and death of bacteria, wherein he attempted to put into proper perspective what the destruction of micro-organisms means. Thus, death cannot be defined by positive criteria; it can be characterised only by the absence of some property which is essential to life, and it is almost

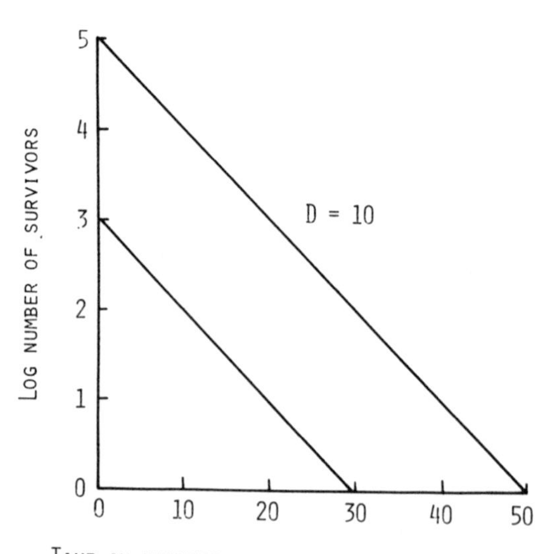


FIG. 1. Hypothetical survivor curves. *D* value is defined as the time in minutes taken to destroy 90% of the viable micro-organisms at some specific temperature.

universally accepted that not all criteria appear at the same time. The loss of reproductive ability might mean micro-organism destruction in one sense, but the cells could well be alive according to some other definition.

From a practical point of view, however, it is reasonable to agree that the destruction of micro-organisms by heat is a first-order reaction, even though variations do appear when destruction at some time–temperature phase does not appear linear; Fig. 1 illustrates a first-order reaction showing the destruction of micro-organisms.

The dairy industry uses heat to destroy micro-organisms both in processing and in sterilisation, and it may be applied in different ways. The most common method uses moist heat, such as boiling water, but other heat forms can be steam or pressurised steam, or autoclaving which is analogous to retorting in canning operations. Occasionally, we may even see people using dry heat, such as the heating in ovens that was once popular in the

TABLE II

RELATIVE SANITISING EFFECTIVENESS OF DIFFERENT HEATING METHODS

| *Type* | *Approximate working temperature (°C)* | *Comments* |
|---|---|---|
| Boiling | 100 | Spores can be viable even after boiling for many hours. Suitable for disinfection but not sterilisation. |
| Free flowing steam | 100 | Not much better than boiling because temperature does not exceed 100 °C, and may be lower at high elevation. |
| Steam under pressure | 120 | Under pressure, temperature better than free flowing steam or boiling water. |
| Hot air | 165 | Spores need about 2 h as protein coagulation needs moist heat. |

dairy industry to sanitise equipment and to sterilise milk bottles. Table II shows the relative effectiveness of the different heating methods. In dealing with the destruction of micro-organisms using heat, it is important to be familiar with the basic terminology employed to describe heat destruction processes. Some common terms to explain what takes place are shown in Figs 1 and 2, but the student is urged to consult a text book on food processing to learn more about $Z$, $F$ (the number of minutes required at a

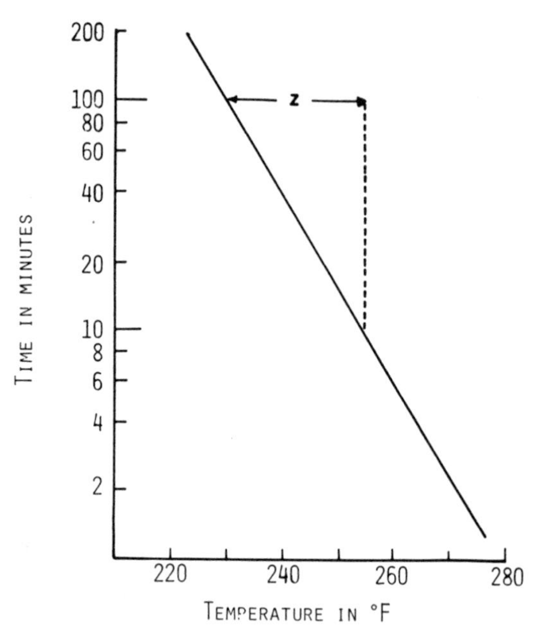


FIG. 2. Hypothetical death curve with its $Z$ value. $Z$ designates the slope of a thermal death curve. It is the number of degrees Fahrenheit required for a specific death curve to reduce the counts 1 log cycle. Here, $Z = 25\,°F$.

specific temperature to destroy a specified number of organisms) and *D* values.

## HEAT TREATMENT OF MILK

From a historical point of view, we should understand that milk was first heated to increase product shelf-life, but later it became mandatory to heat milk to prevent the spread of milk-borne disease. Different people had varied opinions on how milk should be heated, but for many years, milk was pasteurised at 142 °F (61·1 °C) for 30 min in what was termed 'approved equipment'. The organism responsible for tuberculosis was taken to be the test organism, and milk had to be exposed to a sufficient heat and time period to kill this pathogen. Westhoff (1978) published an extensive review about heating milk which should be looked at by students wanting an 'historical update' on this topic. It was not until 1950 that Bell *et al.* (1950) showed that raw milk could transmit Q fever when it was infested with

*Coxiella burnetii* for, when in large numbers, these organisms survive pasteurisation at 143 °F (61·7 °C) for 30 min. In 1950 the United States Public Health Service quietly ordered that the time–temperature requirements for vat pasteurisation of milk be increased to 145 °F (62·8 °C) for 30 min, and 5 °F higher for milks containing sugar.

In a natural way, milk processing technology developed more efficient methods to heat milk, a trend spurred by the suggestion that milk would taste better and last longer if heated to a higher temperature for a shorter time. Vat heating was mostly replaced by high temperature–short time (HTST) heating systems (see Fig. 3). Some other schemes inject culinary quality steam into milk directly, and then draw enough steam vapours off the milk, with vacuum, to compensate for water addition to milk. Each system has special advantages for specific products, but they may also have disadvantages. Some disadvantages are visible, but some problems can be

FIG. 3. The plate heat exchange unit is the central part of a high temperature–short time (HTST) system. Reproduced by courtesy of the APV Co. Ltd, Crawley, Great Britain.

more subtle. Europeans became the leaders in developing ultra-high temperature (UHT) milk which, according to the International Dairy Federation (IDF), means milk is heated to at least 130 °C in a continuous flow for at least 1 s (Burton 1965). The term 'commercially sterile' is applied to milk processed with this type of heating–time arrangement. In the United States, ultra-high temperature pasteurised milk was defined by the Food and Drug Administration (FDA) to mean above 138 °C (280 °F) for at least 2 s, and such milk has to be refrigerated. Table III summarises some current minimum pasteurisation standards.

TABLE III
CURRENT PASTEURISATION SCHEMES

| *Product and method* | *Time–temperature* |
|---|---|
| Vat method | |
| Milk | 30 min/145 °F (62·8 °C) |
| Cream | 30 min/150 °F (65·6 °C) |
| Ice cream mix | 30 min/155 °F (68·3 °C) |
| HTST method | |
| Milk | 15 s/161 °F (71·7 °C) |
| Cream | 15 s/166 °F (74·4 °C) |
| Ice cream mix | 25 s/175 °F (79·4 °C) |
| Ultra-pasteurised | |
| All products | 2 s/280 °F (137·8 °C) |

The student should also consider the fact that there are more novel ways of destroying micro-organisms, with heat being supplied by less conventional sources than those noted so far. Thus, some scientists have been recommending that milk be pasteurised by microwave energy. Jaynes (1975) described a method of heat treating milk with a two-stage regeneration heating system using microwave (2450 MHz) as the energy source. The milk was heated to 72 °C for 15 s, and the results obtained were similar to milk pasteurised by more conventional procedures; Table IV highlights some of the more meaningful data.

While much of the ultra-high heat technology is being used in the European market to produce long-life commercially sterile milk, there is now some effort being put forth in the United States to market UHT milk also. Scientists at the University of Maryland introduced a falling film pasteuriser/steriliser system for sterilising milk. In this method, a thin film of milk is introduced into a sterilising chamber where it comes into contact

TABLE IV

COMPARING THE REDUCTION OF MICRO-ORGANISMS IN MILK HEAT TREATED BY THE MICROWAVE METHOD AND BY A CONVENTIONAL STEAM HEATING TECHNIQUE

| *Counts* | *Raw* | *Microwave* | *Steam* |
|---|---|---|---|
| Standard plate count | $1 \times 10^6$ | $4 \cdot 1 \times 10^2$ | $3 \cdot 5 \times 10^2$ |
| Coliform count | $2 \cdot 9 \times 10^2$ | $<1$ | $<1$ |

with live steam at 280 °F (137·8 °C). The chamber walls are kept cool, during heating, which prevents milk cook-on. Thé milk processed in this system does not have a cooked flavour and appears to have good consumer acceptance; Table V indicates the microbial quality of milks processed by such methods.

The use of heat controls or destroys micro-organisms by inactivating their enzyme systems and/or coagulating cell proteins, and the speed at which organisms are affected varies with the temperature used, and also with the kind of heat being applied. Thus, moist heat is a more effective heating medium than dry heat, and steam under pressure is a better heating system still. The kind and degree of heat used by the dairy industry to control or destroy micro-organisms depends upon the nature of the finished goods; heat can degrade some milk constituents. If one wishes to kill all the living organisms in or on a material, then a sterilisation process is required to accomplish the task. Different methods may be used, but sterilisation

TABLE V

PROCESSED STERILE MILKS STORED AT 28 °C

| *Processing temperature (°C)* | *Colony counts* | |
|---|---|---|
| | *Aerobic* | *Anaerobic* |
| 148·9 | 0 | 0 |
| 143·3 | 0 | 0 |
| 137·8 | 0 | 0 |
| 132·2 | 0 | 0 |
| 126·7 | 25 | 25 |
| 121·1 | 100 | 100 |

Source in part: Westhoff and Doores (1975).

usually means that the material has to be autoclaved at 121 °C (using steam under pressure, 15 lb. in$^{-2}$ brings the temperature in the autoclave to 121 °C) for about 15 min. Such practices are not, however, always possible, so that sterilisation practices vary considerably; for example, the dairy plant staff may sterilise small pieces of equipment by heating them in hot air ovens at 360 °F for 2 h. Less rigorous heat treatments are used in situations where only the pathogenic bacteria need to be eliminated, as for instance, when an operator treats food processing utensils with heat or chemicals, prior to use with milk, to destroy most micro-organisms. In this case, we encounter a term: 'sanitising'. Live steam or hot water can be flushed through milk handling equipment in the same way to destroy most of the organisms left on the surface after a cleaning operation.

## PROCESSING TECHNIQUES TO CONTROL MICRO-ORGANISMS

### Heat Systems

Heat, in its different forms, is used by the dairy industry to control or destroy micro-organisms, and because bacterial cells and spores decrease in a logarithmic fashion when heated, it is possible to plot or calculate a time–temperature curve to project microbial kill. This means that a food processor can partially control the degree of product degradation versus microbial kill that he is willing to accept in order to control microbial activity.

The classical example of using a time–temperature curve to project microbiological kill occurs in the dairy industry when fluid milk is pasteurised at 145 °F (63 °C) for 30 min. This method is called a batch, low-temperature holding (LTH) process, and is carried out in a vat equipped with an appropriate agitator and air-space heater. As you might suspect, a batch method is slow and cumbersome, and more efficient methods are being used in larger factories. A more complex system to kill micro-organisms in less time, but with higher temperatures, is the method of choice. The efficiency of heat transfer equipment depends on its heating or cooling area, its rate of product flow, the thermal conductivity of the metal, and the temperature gradient between the heating or cooling medium and the product. Some different heating systems to pasteurise milk follow.

*Tube and Shell Heaters*

Milk is pumped through the interior of the tubes, while the exterior surface is heated with live steam or hot water.

*Plate Exchange Heater*

Thin stainless steel plates, sealed at the edges with gaskets, are clamped together within a press. The spaces between the plates are alternately filled with milk or a heating medium. The ratio of heating surface to volume is favourable because the milk is in a thin film, and hence the heating operation is efficient (Anon., 1966).

*Direct Steam Injected Heater*

Milk can be heated by injecting culinary quality steam into the liquid. While milk can be heated quickly by such a process, it adds a certain amount of water to the milk and hence the process is almost always combined with a vacuum treatment to flash-off the excess water.

*Steam–Vacuum Heating System*

These systems are finding more use as a pre-heating step for ultra-high heat processes. The process is also being used to improve milk flavour, especially in areas where cattle find their way into garlic or onion flavoured forage.

*Falling Film Evaporators*

These use a heat–vacuum system, but while a conventional evaporator recirculates milk through a heater and then to a liquid separator; the milk in a falling film unit only passes through the tubular heater once in a single pass.

*Swept-Surface Heat Systems*

Heating can be carried out in steam heated 'shell and tube' heaters modified with a swept-surface unit to prevent milk burn-on. All systems strive to heat more quickly, but attempt to avoid burning milk onto the heating surfaces.

**Eradication by Chemicals**

Because all bacteria do not react alike when exposed to chemical biocides or disinfectants, the eradication of micro-organisms by chemical methods can be a complex undertaking, and even a so-called 'over-kill', which may be thought of as using excessive amounts of chemicals to destroy bacteria, may prove unsatisfactory.

The dairy industry, from time to time, shifts its emphasis in controlling the processes of micro-organisms left on utensils or equipment used to handle or process milk and milk products, from heating systems to chemical sanitising agents. It is important, therefore, that technical staff working in this area know that they must use different chemical agents to kill different micro-organisms. In addition, we should realise that a chemical sterilant also changes in its ability to control micro-organisms under varied situations. We know that bacteria, bacterial spores and bacteriophage can be destroyed rather quickly with hypochlorite solutions, even in factories with varied water quality conditions. This fact may not be

TABLE VI

WATER HARDNESS INTERFERES WITH THE EFFICACY OF HYPOCHLORITE AND OTHER SANITISING SOLUTIONS

| *Material* | *Efficiency* |
|---|---|
| Liquid sodium hypochlorite | Good in hard water |
| Powdered sodium hypochlorite | Varies with water hardness |
| Calcium hypochlorite powder | Effective in hard water |
| Iodophors | Adversely affected by hard water |
| Quaternary ammonium compounds | Good to poor, varies with water hardness |

true with other compounds being used to control micro-organisms; Table VI illustrates the effect of water conditions on different chlorinated sanitisers. Interfering substances in a system, such as fats or proteins, dissipate chemical disinfectants and, thus, alter the efficacy of a chemical disinfectant. For example, weak disinfectants, soap being an example, may interfere with a sanitiser to the extent that it could require as much as a 400-fold sanitising concentration above normal concentrations to kill specific kinds of micro-organisms.

*A Review of Chemical Compounds Used to Kill or Control Micro-Organisms*

The need for cleanliness and sanitation can be traced to the beginning of man. One can find passages in the Bible describing rules governing cleanliness and sanitation, as it pertains to the diet, to the care of the diseased and to the handling of wastes. Customs evolved which incorporated these methods of cleaning and sanitising, but it was without any understanding of why the methods worked. At times an observant person would venture an explanation, but, for the most part, these early works were lost in the annals of history. By the 19th century, compounds such as chlorine had become established as being able to prevent foul [illegible]s and to disinfect wounds. Other compounds such as iodine, sulphur and phenol emerged as being of high disinfecting potential. Slowly, the explanations for why the cleaning and sanitising methods worked became understood (Block, 1968).

Lee (1975), a former graduate student of the author, looked at the ability of some 54 different chemical cleaning and sanitising materials to control micro-organisms in membrane systems being used to process milk and whey in the dairy industry. From this work, it was obvious that it is important to look at 'where we are' in respect of the basic technology to control micro-organisms chemically, and a brief review of the topic is presented within the scope of a membrane model (Fig. 4).

FIG. 4. Part of an ultrafiltration system where membranes have been cast in hollow tubes through which whey is filtered to separate whey proteins.

The reverse osmosis/ultrafiltration principle has been adapted to food processing industries in a number of ways (Porter and Michaels, 1970) and the advantages of this principle in applications of this sort are: (1) the processed product suffers little heat damage or phase change; (2) a lower fuel consumption is anticipated in terms of power utilisation and (3) fractionation of a multi-component solution is possible (O'Sullivan, 1971). The recovery of valuable products, such as sugars and proteins, from the waste waters of several food industries has been achieved through ultrafiltration. Most notable is the work of Zall *et al.* (1971) where they demonstrated the concentration and fractionation of cheese whey, both on a pilot plant model (1000 gal. day$^{-1}$) and later at the rate of 30 000 gal. day$^{-1}$ (Zall, 1977). With the advent of membrane processing in the food industry, new problems were encountered, especially with the cleaning and sanitising of the membranes to ensure clean processing equipment.

Today, disinfectants and sanitisers are classified according to their chemical make-up. The major groups are the halogens (chlorine, iodine, fluorine and bromine), heavy metals (silver, copper and mercury), alcohols (ethanol and isopropanol), phenols and bis-phenols, quaternary ammonium compounds, salicylanilides and carbanilides. Each sanitiser group is distinct in its mode of action against micro-organisms, and each reacts differently to the types and ages of micro-organisms encountered. In addition, the effectiveness of each group is governed by specific physical parameters, and these parameters include solvent composition, pH and the presence of electrolytes. Thus, the solvent in which the biocide is carried can either enhance or diminish its biocidal capacity, while the pH of the solution can also govern the effectiveness of the biocide; for instance, weak acids, such as benzoic acid, are dependent on the undissociated ion for the antimicrobial action. The presence of electrolytes can enhance the biocidal effects of organic acids and, yet, is antagonistic toward phenolic compounds. Temperature dependency can be expressed in terms of the energy of activation necessary for disinfection.

The mode of action for the alcohol group can be explained by three types of action: (1) denaturation of proteins; (2) interference with cell metabolism and (3) lytic action (Morton, 1968). Of the various alcohols, methyl alcohol has the weakest bacteriocidal activity. Generally, biocidal activity improves with increase in molecular weight of the alcohol (tertiary alcohols are exceptions) but most research with the alcohol group has centred on the effects of alcohol disinfection on tissues and surface areas of surgical instruments.

Phenol was once widely used as an antiseptic, and the phenol coefficient became the standard against which all other disinfectants were compared. Phenol's main mode of disinfecting action arises from physical damage caused to the cell walls of bacteria or fungi. Bis-phenols are compounds with two phenol groups attached by various types of linkages, and the type and position of the chemical linkage is important in the biocidal ability of the compound. The significance of the chemical structure and types of substitution are well understood. The solubility of these compounds in water is low and, thus, their antimicrobial activity is dependent on their ability to dissolve in water with the aid of solubilising agents. These compounds also lose much of their effectiveness in the presence of organic matter and quaternary ammonium compounds, but stay quite effective in the presence of soap.

Salicylanilides and carbanilides are germicides which have a phenyl carbamide structure in common. They differ in their properties from the bis-phenols, even though they too are widely used in cosmetics and toiletries. These compounds have the unusual ability to retain effectiveness in the presence of skin.

Chlorine is the most widely used of the halogens, and is employed in the purification of water supplies for most cities in the United States. Due to its electron configuration, chlorine is a very strong oxidising agent, reacting quickly with metal ions and organic material but, by so doing, it loses its disinfecting properties. The mechanism of chlorine action has been elucidated by several workers. It is theorised that hypochlorous acid (HOCl) is the actual disinfecting agent, working by combining with the protein of the microbial membrane and thus producing compounds which interfere with cell metabolism. Another theory postulates that chlorine inhibits certain enzymatic reactions which are vital to the micro-organism's function and life. The effectiveness of chlorine is influenced most by the pH of the solution, and this observation concurs with the theory that hypochlorous acid is the disinfecting agent. Other physical factors affecting chlorine's biocidal properties are concentration, temperature, organic matter, water hardness and the presence of ammonia or amino compounds.

Iodine, another halogen, is like chlorine in its reactive nature. It is the free iodine, however, which acts as the biocidal agent causing cell protein precipitation. Iodine's advantages as a germicide can be summarised as: (1) effective over a wide pH range; (2) concentration necessary for disinfection varies little with types of micro-organism; (3) low toxicity; (4) fast action and (5) its colour is an index of strength (Rudolph and Levine, 1941).

Another widely used sanitiser group is the quaternary ammonium compounds which were first introduced in 1935. 'Quats', as these compounds are frequently called, have the distinction of being: (1) non-corrosive, as they do not contain phenols, iodine or active chlorine; (2) non-odorous; (3) highly stable; (4) non-colouring and (5) effective against many types of micro-organism. 'Quats', however, are incompatible with a long list of common compounds, most notably soap, cations in hard water and organic matter. An excellent monograph prepared by Lawrence (1950) covers practically all aspects associated with the 'quats'.

Methods for the testing and analysis of the germicidal powers of a compound were established and based on: (1) time of exposure; (2) temperature of exposure and (3) age and type of culture, and the effects were compared to that of a universal standard. The Rideal–Walker method, introduced in 1903, using phenol as the standard and nutrient broth as the growth medium, became the standard bacteriological test. Even today, in some parts of the world, the same procedure outlined in 1903 is being used. Slight modifications to the procedure have been introduced by other workers since then, amongst which is the Official Food and Drug Administration Method (Rueble and Brewer, 1931) for the United States. This method, in turn, has been superseded by the AOAC (Association of Official Analytical Chemists) procedure which allows for a choice of three sub-culture media. In addition, procedures were developed which specifically test for surface disinfecting abilities, antifungal properties and a specific test for quaternary ammonium compounds.

On a practical level, the measurement of death and/or survival of micro-organisms is the ability of the micro-organism to reproduce when provided with favourable growing conditions. In the case of reverse osmosis/ultrafiltration membranes, a very suitable environment for microbial growth is provided by membranes that are inadequately cleaned, so leaving some food residuals behind. This microbial contamination of the membrane surfaces, especially with the cellulose acetate types, can degrade membranes drastically.

The function of detergents is mainly to loosen and remove dirt and grease which have become attached to a surface. Non-ionic detergents, which have found wide usage in industry, are stable over a wide temperature and pH range, but these detergents do not possess any biostatic or biocidal properties *per se*. Anionic detergents however, are less effective cleaners than the non-ionics, but they do possess some biostatic ability, especially toward Gram-positive organisms. Cationic detergents are best represented by the quaternary ammonium compounds, and these are clearly biocidal as well. Soaps, which are no longer used for cleaning equipment in industry, do possess some biocidal and biostatic properties, but the efficiency of killing is dependent on the type and proportion of fatty acids. Generally, capric acid is more effective than lauric. Resistance of the micro-organisms to soap depends on species and strain and in addition, factors such as pH, temperature and organic matter are important in terms of effectiveness.

The practice of recycling cleaners and sanitisers has been studied to determine whether cost savings can be accomplished without jeopardising cleaning and sanitising effectiveness. There are basically two objections raised against the recycling of cleaners and sanitisers, namely, the increase of food residuals in the solution, and the increase of micro-organisms in the solution. Work by Maxie (1964) indicates that recycling of cleaners and sanitisers is feasible from a public health point of view, and Zall and Brown (1976) demonstrated that a recycling method for cleaning milking equipment was feasible too.

The application of our knowledge about germicides and detergents for a specific function is indeed a challenging one, for in addition to the obvious use in the medical field, germicides are important in both industry and the home. Industrial applications have been developed to meet the needs of a specific process, and both the product being processed and the micro-organisms encountered in this process have determined the methods of sanitising that can be used. The end result of this development is that, today, the cleaning and sanitising of equipment used in the handling of human food is explicitly described by rules set forth by National, State and Local Governments.

Many different cleaners, sanitisers and detergent/sanitiser combinations are available for industry's use, but the development of special procedures, such as those employed for the cleaning of reverse osmosis/ultrafiltration membranes fouled by organic matter from brackish water, is a useful example of problems of application. Thus, investigators found that cleaning the membranes with an enzyme detergent was most successful in restoring flux and, in addition, that sodium perborate, EDTA or calcium hypochlorite treatment seemed equally able to rejuvenate membrane flux rates. However, although hypochlorites were the commonly recommended biocide for sanitising membranes before a product run, it became evident that cellulose acetate membranes were susceptible to degradation by chlorine. Nevertheless, very little work was done in trying to find non-corrosive sanitisers for cellulose acetate membranes, and part of the difficulty lies with the fact that the sanitiser must not be rejected by the membrane due to molecular size, charge or structure, otherwise the permeate side of the membranes are not sanitised. Instead, work was directed toward the manufacture of non-cellulosic membranes which would be more resistant to pH, temperature and existing sanitisers, and membrane manufacturers, in general, have shown little expertise in past years in being able to recommend good procedures for cleaning and sanitising their hardware; the usual response was merely a few warnings concerning temperature, pH and sanitiser concentration.

Water is a medium upon which all life functions, and yet the quality of this water varies with location and seasons of the year. This variation in water quality affects the performance of many processes which utilise water. Laundry is greyer because of the hardness of the water; pipes corrode because of acid in the water; and the odour and taste of water may be peculiar due to gases dissolved in it. Water quality, likewise, is an important consideration in terms of cleaner/sanitiser performance (the effects of hard water on the biocidal activities of some sanitisers, especially quaternary ammonium compounds, are well known) as well as in respect of membrane performance and life. Thus, it is reported that calcium sulphate ($CaSO_4$) precipitates can decrease membrane flux rates, and that other divalent ions found in water can contribute to scaling and fouling problems. Fine rust particles from the water will deposit on membrane surfaces and adhere so tightly that removal is almost impossible, while magnesium and calcium ions can combine with sulphate, carbonate and phosphate ions to form precipitates on membrane surfaces. Hence, membrane manufacturers recommend the use of softened water when the total hardness in the municipal water supply is greater than 50 ppm (parts per million).

Total water hardness is measured by the amount of calcium and magnesium ions in the water. In procedures outlined by Standard Methods (APHA, 1975), one can readily measure the amount of these ions in water and, in addition, convenient chemical kits are available for field testing of not only water hardness, but also chlorine, salt, dissolved oxygen, etc. Once the decision is made to 'soften' the water, that is, to remove the calcium and magnesium ions or other specific ions, several methods of softening are available. The principle involved is ion exchange, in which the calcium and magnesium ions are exchanged for sodium ions. On the laboratory level, more sophisticated resins are used to remove not only metal cations, but also anions and organic groups, such as proteins, peptides, amino acids and nitrates. These deionisers pass out very high quality water, but the components from the water, which are then trapped in the resin beds, can serve as food for any micro-organisms which are fed into the unit by water. Hence, consideration must be given to rejuvenating fouled resin beds by sterilising them with formaldehyde or chlorinated compounds; otherwise, a contaminated deionised water will result.

The use of sequestering agents to form metal complexes has been known for many years. This discovery came about with the recognition that each metal has a co-ordination number which it exhibits, and that the metal complex which it forms can exist in isomeric forms. Ethylenediamine, one of the first chelating agents to be studied, was shown to bind to platinum in a ring-type structure. As a complex, the chelated compound is inherently more stable than closely related non-chelated compounds, and of the chelating agents available, ethylenediamine-triacetic acid (EDTA) and citric acid are most frequently used by the food and membrane industries.

**Using Refrigerated Systems to Control Microbial Growth**
As we make changes in the production, processing, and distribution of milk and milk products, we solve some problems and apparently create new difficulties. Modern dairy farms have automated milking systems together with on-farm refrigerated bulk storage tanks, and these innovations seem to have changed the quality of raw milk by altering its microflora. The shift in kinds of micro-organisms has been in favour of psychrotrophic bacteria with a decrease in streptococci (souring species); i.e. those micro-organisms that can grow and multiply at 7 °C or less. Researchers believe that these psychrotrophic bacteria are responsible for causing poor flavours in milk, for producing low cheese yields, because many of them are proteolytic, and for causing refrigerated dairy products to spoil sooner than they might if the micro-organisms were not present. We have known for many years that some micro-organisms can multiply in cold milk, but for a long time, the milk industry got along reasonably well by obeying the cardinal rules of keeping milk *clean* and *cold.* However, with the introduction of bulk handling, along with refrigerated milk storage, defects in milk quality began showing up in different milk marketing areas.

## MILK QUALITY PROBLEMS IN THE UNITED STATES

Whether right or wrong, the American consumer expects pasteurised milk to remain good about two weeks from the time of purchase. Thus, if we think about the real age of market milk in America, we must recognise that raw milk stays on the producer's farm for at least two days, and then it may take a further day or two for the milk to be transported to a city plant for processing. After pasteurisation of the milk in the factory, the packaged goods (milk in plastic cartons) may then spend one to two days in a food shop before multi-packages are bought and taken to the customer's home. As it is common practice for a week's supply to be purchased at one time, some milk could easily be 14 days and older when it is finally consumed, and Americans go to a lot of trouble trying to market fluid milk within this time framework. Nevertheless, researchers in the United States found that about 10 % of the market milk in a mid-west major city had poor flavour and similar information was reported by colleagues in my own university (Cornell) when they reported that 9·6 % of the milk in New York State is of poor quality.

## MILK QUALITY PROBLEMS IN EUROPE

Milk in Europe is also subject to contamination with psychrotrophs, although rapid distribution to the consumer (2–3 days as against 14 days in America) has tended to minimise the problem. In particular, Thomas and Thomas (1973) have published widely on the bacterial contamination of milking systems, and have also reviewed the literature on psychrotrophic bacteria in milk as recently as 1978. It is clear, therefore, that British scientists, as well as other world experts, are speaking out against the fact that much of today's milk is oxidised and rancid. The problems associated with describing specific areas of taste are difficult, especially when not everyone is a trained food taster, nor can we quantify the subtle differences in taste. We still have to use parameters like acid degree values, psychrotrophic bacterial counts, pH, titratable acidity and, of course, our tongues. The use of taste to grade milk is important, because the science of taste is still mainly an art.

These defects are caused by people using less than ideal milk handling practices during production, processing and in distributing dairy products. It is also obvious that current quality control measures are not coping with some of the problems now coming to light.

## CHANGES NEEDED TO IMPROVE MILK HARVESTING EQUIPMENT

It is common knowledge in some areas (both in the USA and in Europe) that many dairy farmers do not clean milking equipment after each use. Some people only rinse their milking hardware after night milking, and feel that it is satisfactory to wash the milk handling equipment with detergents only once a day. Such an attitude prevailed in past years, but some of the practices have been carried over into larger and more complex operations, including when dairymen switched from the can milk system to bulk milk.

An automated milking centre on a modern farm is a complicated system (Fig. 5). We find glass lines, stainless steel tubing, plastic and rubber hoses, milking claws, and different sorts of gadgetry inter-connected one to the other with different joining mechanisms. In fact, air, cleaners, sanitisers and acids are often mixed in different solutions of varied temperatures, and then pumped through common lines which do not always drain free. The work of Zall and Brown (1976) during a 1973–1976 period pointed out some of the problems in this area, and they suggested some changes which would

FIG. 5. Some of the 'hardware' currently used in the milking parlours of large dairy farms to harvest milk from many cows simultaneously. Reproduced by courtesy of Alfa-Laval, Cwmbran, Great Britain.

improve the situation. Several publications were released that described the advantages of using recycling cleaning chemicals, and showed how dairymen could reduce cleaning costs and improve cleaning. The concept of recycling cleaning materials on farms may be important enough to motivate some people to wash milk handling equipment more often. In fact, it might be best for the dairy industry to re-examine the whole spectrum of milk handling equipment as it used to produce, collect, transport and process milk from the farm to the consumer. Each unit operation contributes positively or negatively to milk quality.

## USING AGED MILK IN CULTURED PRODUCTS

Most defects in cultured products probably relate to poor milk, undesirable starter and/or a breakdown in sanitation; we could also include inadequate workmanship, and a weak or poorly thought out quality assurance programme. For example, some of us believe that a breakdown of milk fat to cause rancidity is associated with the growth of bacteria as well as with biochemical degradation, and some researchers speculate that high counts of proteolytic psychrotrophic bacteria in milk for cheesemaking can result in enough casein degradation to cause vat failures. If cheese is made from such milk, then the casein particles are so damaged as to make cheesemaking difficult, and they suspect that the casein micelles are altered to the extent that a firm curd cannot be formed.

At this time, there are no reliable data to deal with the effects of psychrotrophs on product yields of Cheddar cheese. There is, however, some information about yields of cottage cheese made with milk containing large numbers of psychrotrophs. University researchers published work showing that when cottage cheese was made with milk contaminated with psychrotrophs it lost about 0·5% in yield even without vat failures. In addition, when psychrotrophic organisms were isolated from milk obtained from a cheese plant having operational problems, and these bacteria were inoculated into milk used for making cottage cheese, then yields again dropped by a measurable amount.

Cheesemakers have always complained about cheese quality when it was manufactured from milk stored too long prior to pasteurisation, but the bacteriological problems in today's milk are different from those they had to live with a few years ago. Thus, milk no longer sours due to lactic acid bacteria, but the supply will now be degraded or spoiled by cold storage micro-organisms growing at 5°C, which work on fat and protein to break down these constituents to produce more subtle defects.

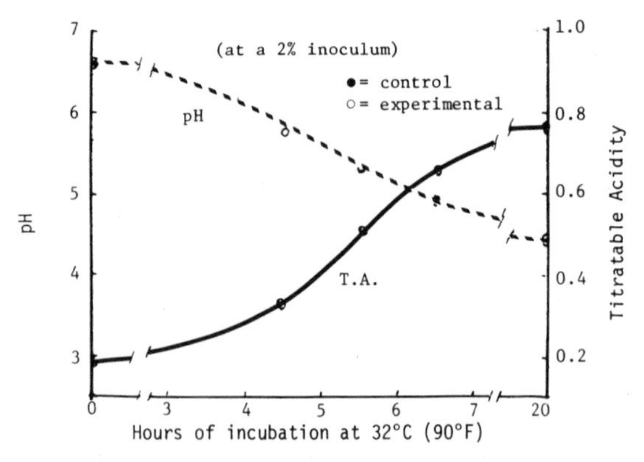


FIG. 6. Comparing the gross fermentation characteristics of 2-day old raw milk with 10-day old farm-heated (165°F/10 s) milk inoculated with a 2% v/v buttermilk culture. (Note overlay of data points.)

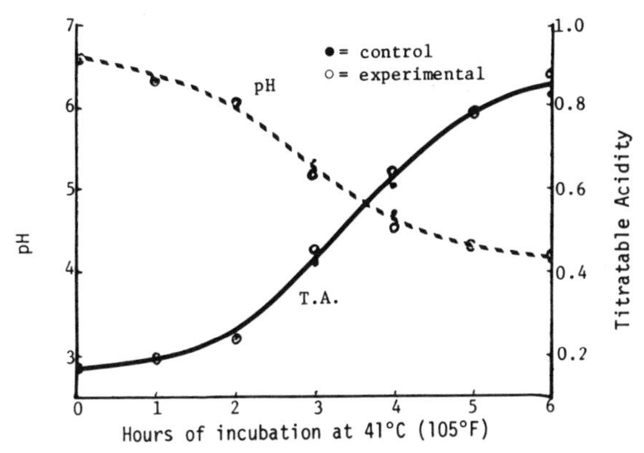


FIG. 7. Comparing the gross fermentation characteristics of 2-day old raw milk with farm-heated (165 °F/10 s) milk inoculated with a 2 % v/v yoghurt culture. (Note overlay of data points.)

To offset this problem, it appears that heating milk as it is produced might remedy the situation. Zall (1979) suggested that stored, farm-heated milk reacts favourably to different culturing situations, and Figs 6 and 7 show that culture activity in fresh bulk milk (2-day old) and 10-day old milk is not dissimilar. The yoghurt culture was run along with the buttermilk starter because yoghurt processing tends to be more sensitive to inhibitory materials in milk than cultures for cheesemaking, at least in the early stages. It is logical to believe that some phases in cheesemaking could be controlled better by heat treating milk on the firms, and cultured products, like buttermilk and yoghurt, can be made about as well with aged, heated farm milk as with fresh milk.

## COOLING TO CONTROL GROWTH

While we have touched somewhat on the way that heat and chemicals can be used to inactivate micro-organisms in milk, we should consider controlling micro-organisms in milk by cooling and refrigeration, not forgetting the ever-present problem of psychrotrophic organisms.

Milk markets were broadened enormously by using refrigeration to keep milk 'sweet' for longer periods, so that whereas milk is still being delivered more than once daily in some countries because it spoils rapidly by souring, the more developed countries use refrigeration vending systems to keep the milk fresh. Thus, because most micro-organisms reproduce more slowly in

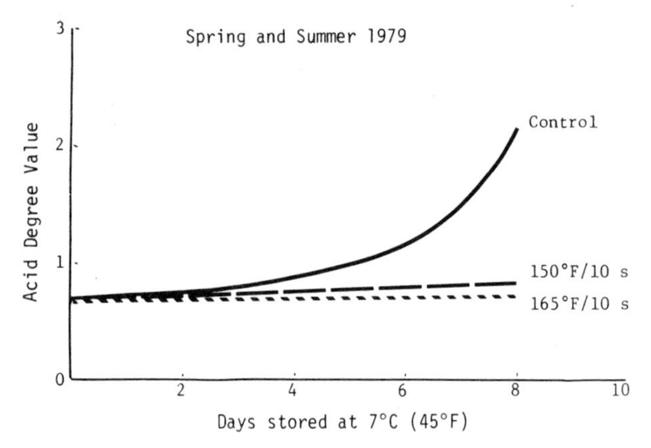


FIG. 8. Acid degree values of unheated and farm-heated milk during storage. (Spring and summer, 1979.)

colder environments, the deterioration of milk is lessened both bacteriologically and chemically when milk is cooled. Chemically, the rates of reactions depend on temperature, and they will be slower in cold milk than in warmer products. Enzymes are especially sensitive to temperature and, while not destroyed on cooling, they react at slow rates in cold milk systems. However, some of the hydrolytic enzymes which are known to be excreted by psychrotrophic organisms continue to function at 0–8 °C, and still others cause problems even at 'freezer' temperatures of −26 °C.

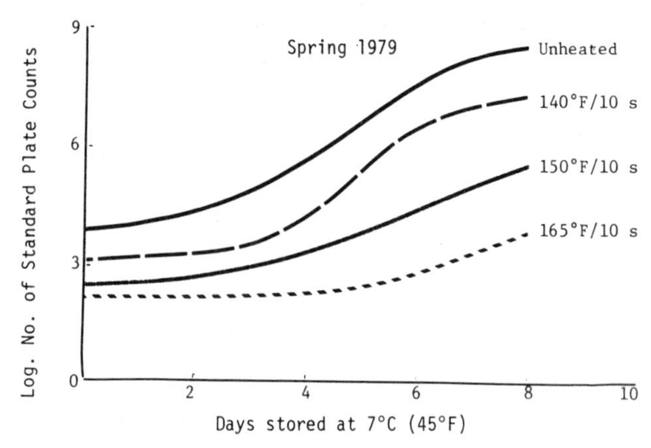


FIG. 9. Bacterial counts of farm-heated milk during storage when heated at different time–temperature combinations. (Spring, 1979.)

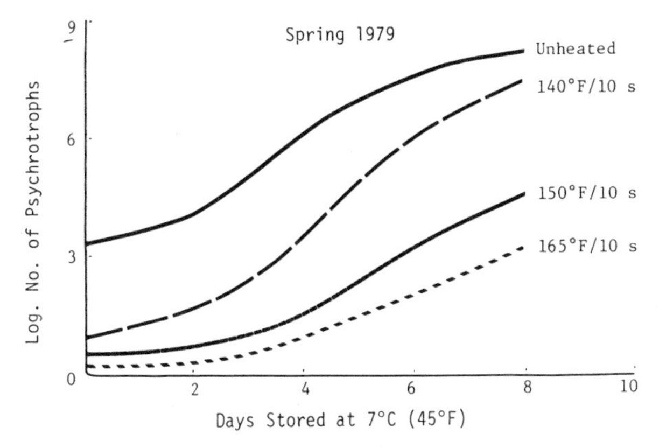


FIG. 10. Psychrotroph counts of farm-heated milk during storage when heated at different time–temperature combinations. (Spring, 1979.)

Zall (1979) reviewed the idea of controlling micro-organisms in milk by both cold milk storage and by the combined techniques of heat treating milk (sub-pasteurisation) on farms prior to storing it in refrigerated bulk tanks. The data are summarised in Figs 8, 9 and 10.

## USING LOW TEMPERATURES TO CONTROL MICRO-ORGANISMS

We have already stated that as temperatures drop below the optimum point for micro-organisms, metabolic activity is slowed but not necessarily stopped. Some bacteria, as well as moulds and yeasts, will grow below 0 °C; for example, moulds like *Cladosporium* and *Penicillium* grow as low as −7·5 °C, and bacteria can grow in the extreme cold salt water of the Arctic region. However, as water freezes, the concentration of solids changes, so much so that the altered characteristics may retard or kill the microbial flora. The process does, therefore, destroy large numbers of micro-organisms in milk and other foodstuffs, and at least 50 % or more of the organisms found to be viable in milk will be destroyed when milk is frozen. Obviously, a micro-organism's ability to survive freezing depends upon its type, its state and the freezing temperature used. Some factors one might wish to consider when dealing with the lethal effects of freezing and sub-freezing temperatures might be: (1) the kind of organisms and their physiological state; (2) temperature during freezing and storage; (3) length of storage; (4) food type and (5) alternate freezing and thawing.

## GAS AND RELATED SUBSTANCES

Killing micro-organisms with gas is a function of gas concentration, time of exposure and temperature of exposure, and the effect on micro-organisms pretty much follows the rules set by other physio-chemical techniques discussed in this chapter.

Ethylene oxide is a gaseous sterilising agent widely used in different areas of the food industry to control or destroy unwanted microbial contaminants in foodstuffs. It had its birth in the food industry, and continues to play an important role in the destruction of thermophiles, yeasts and moulds in many of the condiments used in the formulation of flavoured dairy foods. As to using this gaseous agent in dairy products, it is understood that high concentrations of the gas can degrade certain amino acids in casein and, thus, it is not used to control micro-organisms in milk. There are other reasons too, but overall, the technique of gaseous sterilisation ought not to be discounted.

Chemical sterilisation of milk is permitted in different parts of the world. Siegenthaler (1965) reported that the hydrogen peroxide treatment of milk destroyed 99–100% of micrococci and all of the propionic bacteria and coliforms. He is an avid supporter of the use of hydrogen peroxide in milk, having done outstanding work with the material in different developing countries throughout the world. Unfortunately, the spores of anaerobes and aerobes survive the hydrogen peroxide treatment. The United States permits the use of hydrogen peroxide as a sterilising agent in cheesemilk, but with a usage not to exceed 0·05% $H_2O_2$. Peroxide is, perhaps, used most often to sterilise packaging materials used to contain dairy products. Flat paper or rolls can be sent through solutions of hydrogen peroxide, and can then be sent to a heated chamber to drive off the gas.

Different applications for gas are being found in the industry, and others will be coming on the market as the technology of applying gaseous substances evolves further. One commonly used gaseous substance is chlorine, and this substance is regularly used in cheese plants throughout the world. It is common practice to find people spraying mists of chlorine solutions of 200 ppm, much like a gas, into cheesemaking/processing rooms, and it will also be used in packaging areas. While the use of this technique is valuable in controlling atmospheric moulds in factories, one must wonder what harm this practice might do over a period of time to creamery workers. We know that chlorine and other halogens are corrosive and cause damage to metal machinery in food plants and, as the practice is not without danger, this author must frown upon its widespread use.

## CLEANING MILK HANDLING EQUIPMENT TO CONTROL MICROBIAL GROWTH

The most obvious way to reduce microbial contamination in the milk supply would be to insist that milk handling equipment be scrupulously clean. From a microbiological point of view, this seems obvious, and one must wonder why the issue need even be considered. However, as we have already learned, some of our milk harvesting equipment needs engineering changes to improve 'cleanability' (see Fig. 11) and we also know that milk handlers (people engaged in the production, processing and distribution of milk) need continued education and motivation to treat milk better. As to the cleaning of milk handling equipment, there are some basic methods to

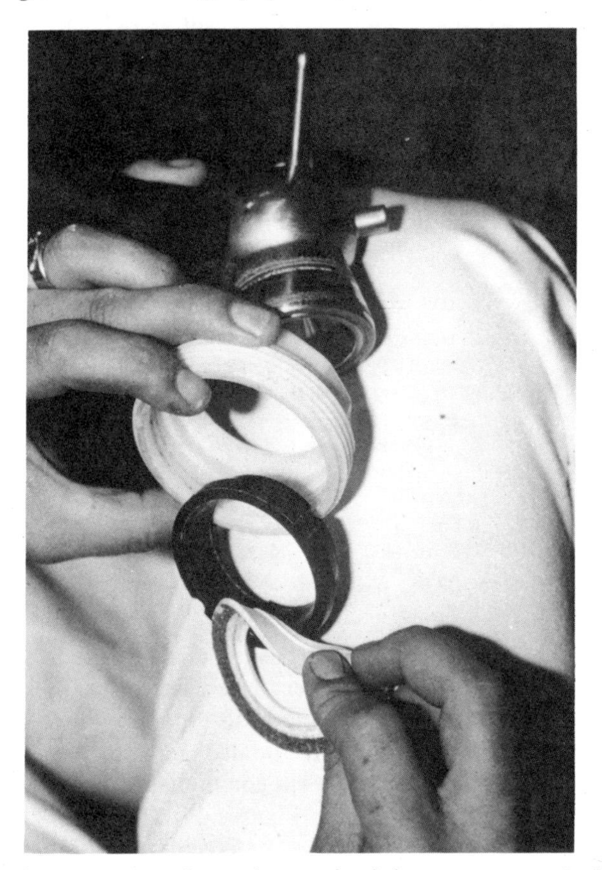

FIG. 11. Sometimes the gaskets and connecting fittings are not properly cleaned.

do the job best and the dairy scientist can look at cleaning techniques as follows.

The cleaning and sanitising of all milk contact surfaces is essential to prevent microbial contamination of the milk, and this caution is not only important from a milk quality point of view, but it also is a matter of importance to public health. Obviously, the chemistry of cleaning is somewhat complex, but it is not so mysterious that we cannot sum up parts of the art for our own use.

The purpose of cleaning equipment is to remove soil substances, and this act is carried out by cleaning the contact surfaces with water and a cleaner or detergent. The activity, when done properly, renders the surfaces almost free of bacteria, and they can then be treated with a sanitiser to make the equipment commercially sanitary. A detergent is a chemical compound that is added to water to increase the ability of water to remove soils. Soil is a term used to mean material which has to be removed in a cleaning operation, including milk residue, water deposits, cleaner and sanitiser residues, dust, sediment, or any other foreign material left on equipment surfaces. The sanitiser can be a physical or chemical agent that is capable of destroying most of the micro-organisms left on a surface after washing; these agents include steam, hot water, chlorine, etc. The chemical

TABLE VII

GROSS CHEMICAL COMPOSITION, ON A DRY BASIS, OF MILK SOILS OBTAINED FROM COLD SURFACES

| *Composition* | *Soil* (%) |
|---|---|
| Lactose | 38 |
| Fat | 30 |
| Proteins | 26 |
| Ash | 6 |

compositions of some milk soils have been studied in some detail and Table VII summarises some of the more recent data. The residues found on heated surfaces of machinery varied according to type of surface and kind of machine. In general, hot surfaces precipitated more proteins and tended to produce milkstone-like substances. The common methods of treating these soils are presented in Table VIII. Figures 12, 13 and 14 illustrate the relationship between soils removal and detergent concentration, exposure, time and temperature.

TABLE VIII

COMMON METHODS USED TO TREAT MILK SOILS

| *Material to remove* | *Cleaning agent* | *Product of reaction* | *Conditions required* |
|---|---|---|---|
| General soil | Water | Dissolved soil and suspended soil | Agitation |
| Fat | Alkaline detergent | Soap and emulsified fat | Agitation |
| Protein | Alkaline detergent or acid detergent | Dissolved protein and suspended solids | Agitation |
| Mineral film | Acid detergent | Mineral salt and acid | Agitation |
| Bacteria | Sanitiser | Bacteria-free surface | Contact time, strength, temperature |

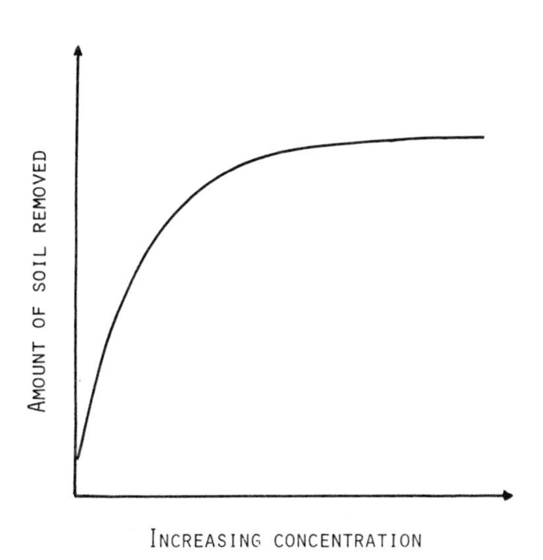


FIG. 12. Adding increasing amounts of cleaning material to washing solutions will not necessarily remove more soil substances.

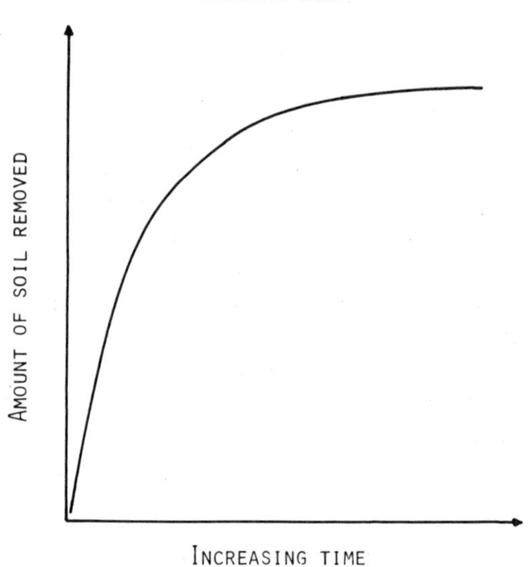


FIG. 13. Increasing cleaning time does not necessarily remove additional soil material.

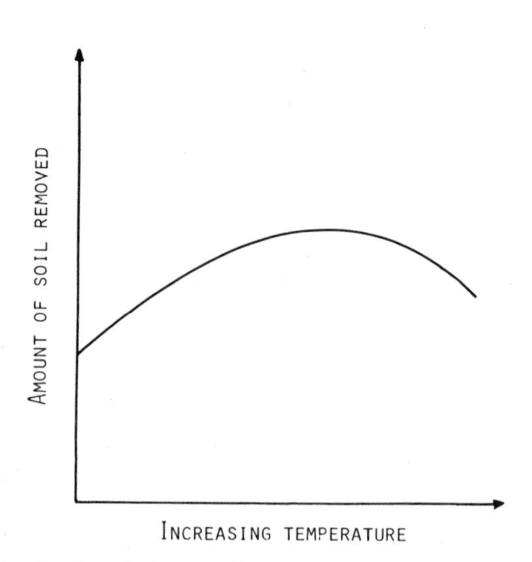


FIG. 14. Heating cleaning solutions too high may fix soils more firmly onto milk processing equipment.

**What do we Mean by a Surface Being Cleaned?**

'Clean' must be interpreted with judgement because the term depends upon too many factors. When a dairy technologist or quality control person deals with cleanliness, the emphasis is most often placed upon the facts or causes of uncleanliness. It is important to be able to distinguish between unsightliness, and factors more important to consumer health needs (Lahr, 1973). Some techniques for maintaining cleanliness to control micro-organisms follow.

*Inspection by Sight*

This method may be used to detect visible film, scale or deposit. A surface uniformly bright and shiny can be presumed to be clean. Adequate lighting should now be available in all milk houses.

If the appearance of stainless steel is dull, it may still be clean if the condition is due to wear and chemical action (e.g. $H_2S$), or the dullness may be due to a film of scale material. In the latter case, the application of an acid milkstone remover will remove the film and leave a bright surface. Farm utensils frequently respond favourably to such treatment.

*Inspection by Feel*

Fingers can detect soil residues.

*Odour*

The milk should not smell 'metallic nor milky'. The presence of any odour at all is proof of contamination or oxidation.

*Swab Tests*

Using a dry cloth, these tests are valuable. Water-scale or milk deposit will appear whitish on black cloth. Rust or metallic oxides will show best on white cloth.

*Bacterial Counts*

As described by approved current procedures, bacterial counts are valuable because they are direct evidence of the quality of sanitising. For instance, psychrotrophic bacteria are present when sanitation practices are poor.

*Freedom from Acids, Alkalis and Detergents*

If this is to be determined, the equipment to be tested must be rinsed with water as in sanitising, or else any accumulated drainage water must be collected. To separate, equal amounts (10 ml) of the rinsing or drainage water and the normal plant water (control), 0·1 ml of 0·04% Brom Thymol Blue indicator is added. If the test sample turns yellowish, then acid traces are present. If blue is developed or deepened, then alkali traces are left behind. If the drainage on shaking produces suds, the carry-over of considerable traces of a synthetic detergent, present either in the cleanser or sanitiser or both, is probably responsible; correction lies in better rinsing.

*Black Lamp*

The use of the 'black lamp' (ultra-violet rays filtered) has been found useful in determining soil on clean metal surfaces. Metallic surfaces free from oxides, sulphides, etc., will not absorb this form of energy, but many substances, notably inorganic substances like calcium salts and organics such as casein, will absorb this energy and give off light. New light-weight equipment is available for use by plant workers to check for clean conditions, and this, and other tools, should be part of any inspector's quality control equipment.

**Water Quality is Important to Good Cleaning**

The major component in cleaning solutions is, of course, water and yet this commodity fails to get the proper kind of attention it needs to maximise cleaning efficacy. Water dissolves soil material and carries the cleaning compounds to the surfaces of equipment being cleaned, but it is important that most waters are not pure, and contain substances which can interfere with cleaning schemes. Thus, water hardness reacts with most cleaning chemicals, and the salts of magnesium and calcium are major problems. In fact, it is the deposition of these salts upon equipment surfaces during rinsing operations that causes water spots, upon which micro-organisms set up 'housekeeping'. Cleaning programmes must, therefore, be adjusted to deal with any water hardness that may neutralise or dissipate the ability of detergents and sanitisers.

Hardness is classified as either temporary or permanent. Temporary hardness is caused by calcium or magnesium bicarbonate, and these can be precipitated by heat. These impurities can also be precipitated by many alkaline materials, and are often left behind on plant surfaces as scum. Permanent hardness is mostly caused by sulphates and chlorides of calcium and magnesium. These, too, can be precipitated from water, but not by heat. A rule of thumb used to be that a milk plant should soften its water when the water supply had a water hardness greater than six grains per gallon ($\equiv$112 ppm).

Water hardness in the United States is classified as shown in Table IX. Soft and moderately hard water can be used for cleaning milk handling machinery, but the cleaner selected ought to contain a water conditioning material within its formulation. Hard water really requires conditioning by ion exchange systems, and final rinse waters may need an acid treatment. In addition, neutralised detergents show a decrease in 'rinsability' and this, in turn, creates a problem of residual films being left on equipment. These films, along with water 'spots', provide the stage upon which organic mass can become attached and, on these films, microbial populations also thrive.

TABLE IX
RELATIVE HARDNESS OF WATERS IN THE USA

| | *Grains per gallon* | *Parts per million* |
|---|---|---|
| Soft | 0–3·5 | 0–60 |
| Moderately hard | 3·5–7·0 | 60–120 |
| Hard | 7·0–10·5 | 120–180 |
| Very hard | >10·5 | >180 |

It is prudent to suggest that experts in the field be consulted on how best to deal with any difficult water supplies that have to be used in milk factories.

There are many good textbooks in print, as well as dairy industry bulletins, that cover the chemistry of cleaning materials, and this brief treatment of the importance of cleaning is meant to alert readers to the fact that micro-organisms in milk need to be controlled by good cleaning practices. The dairy microbiologist will do his/her profession a disservice if he or she fails to consider the causes of microbial contamination of a milk supply. It is not enough to know the kinds of micro-organisms one finds and identifies in milk, but rather to use such knowledge to indicate why or how they might have found their way into the milk supply. By knowing this, a dairy scientist can better deal with the control and destruction of micro-organisms.

## BACTERIOLOGICAL EVALUATION OF PROBLEM AREAS IN MILK HANDLING

The ideal locations to obtain samples for bacteriological analysis are at natural breaks in the system. Such areas are storage tanks, balance tanks, filler bowls and filled product containers; sampling valves and rubber sampling plugs provide another area of possible contamination. Rinse counts, contact plates and swab tests may be used for determining the cleanliness of milk contact surfaces and, in most cases, they are necessary to determine the source of a problem. Thus, samples taken from various locations in processing systems will usually indicate the area of contamination, and the results of holding quality tests should indicate potential shelf-life. Flavour evaluation should not be ignored, especially of samples held for shelf-life tests, for conditions, such as 'lacks freshness', putrid and spoiled tastes, usually indicate bacterial contamination after pasteurisation.

## VISUAL INSPECTION

Those milk contact surfaces which are the most difficult to clean should be checked visually and such items include pumps, return lines, valves and vertical discharge pipes. They can provide an indication of the cleanliness of the entire system. When possible, disassembly of parts in a system should be done after the surfaces are dry. If there are any unclean surfaces, they should be pointed out to processing or cleaning personnel, and frequent processing line checks should be made of critical areas; written reports of the findings can be posted. The psychological effect of routine monitoring on personnel who know they are being checked can be very valuable.

A black light can also be used to check in areas which are difficult to clean, and this technique is the most applicable to large surface areas such as storage and transport tanks. The colour of deposits usually provides some indication as to the composition of the film, and the cause. A dullness or white deposit on stainless steel usually indicates hard water or alkaline deposits. A bluish or rainbow hue is frequently an indication of a protein film.

## SANITISATION

### Sanitisation v. Sterilisation

The cleaning of equipment is intended to remove all food residues and foreign matter from contact surfaces, but this sanitation step does not guarantee a sanitary surface at the time of the next use. For this purpose an efficient bacteriocidal treatment is necessary, and sanitisation, rather than sterilisation (a more rigorous and difficult procedure) is the objective of the bacteriocidal treatment.

The United States Public Health Service (USPHS) Grade 'A' pasteurised milk ordinance has offered this definition: 'Sanitisation is the application of any effective method or substance to a clean surface for the destruction of pathogens and of other organisms as far as is practicable. Such treatment shall not adversely affect the equipment, the milk or milk product, or the health of consumers, and shall be acceptable to the health authority'.

By contrast, sterilisation is a treatment or process which destroys all micro-organisms including spores, and requires much higher temperatures than are generally feasible in fluid milk bottling plants. For example, 121 °C (250 °F) for not less than 15 min is the required treatment to sterilise laboratory glass-ware and metal equipment; however, this cannot be done practically in the factory. Consequently, sanitising is the common plant practice used in industry rather than sterilisation.

For the successful use of any sanitising agent, the equipment surfaces to be sanitised must be absolutely free of organic matter (fat, protein and/or milkstone films). It is impossible to properly sanitise equipment with chemicals unless it has been cleaned and is void of any milk residues, otherwise the sanitiser will only be effective against organisms on the surface of the soil. Similarly when an inadequate hot water or steam treatment is used, the heat may bake on the soil and yet be insufficient to penetrate through the soil to kill all the bacteria, especially spore-formers. Hence, the next milk to flow over the surface will be contaminated. Proper heat treatment, however, has the advantage of deep penetration into equipment joints and all surfaces, killing bacteria in areas where chemicals may fail to penetrate. In addition, sanitisation processes are best employed just prior to use of the equipment so that any surviving organisms will not have time to multiply and recontaminate surfaces.

## Types of Sanitisation

### *Heat*

*Hot water.* Hot water is an excellent agent for sanitising either pasteurised or raw product contact surfaces. Totally enclosed, confined areas are easiest to sanitise with hot water, but even pasteuriser surge tanks can be sanitised with hot water by spray ball or other distributing device. Many milk plants confronted with a difficult coliform contamination problem have found it helpful to resort to heat, and the heat transmission properties of metal equipment assure complete sanitisation.

*Steam.* Steam is not recommended for sanitising because, as commonly used, it causes: (1) heat stresses which may crack soldered seams and welds, especially in stainless steel equipment; (2) a waste of energy as steam is dissipated to the atmosphere; (3) leaky valves, and the rapid deterioration of rubber hoses; (4) noise and (5) the destruction of paint on walls and equipment.

*Chemical*

*Hypochlorites.* The most common type of chlorine sanitisers used in the dairy industry are hypochlorites. They are economical and effective for plant use. Sodium or calcium hypochlorites at varying strengths may be purchased in either granular or liquid form, and sodium hypochlorite is also available from on-site generators using common salt, water and electricity; the lower pH of on-site generated hypochlorite offers an equivalent bacterial kill at lower concentrations. Chlorine in the undiluted form can be hazardous and corrosive and care should be taken to prepare proper strengths and to prevent injury and damage to equipment.

*Elemental chlorine.* Chlorine is available as a gas in cylinders.

*Organic chlorine compounds.* These compounds, such as Chloramine-T, are significantly affected by pH. Chloramine-T is much slower acting than the inorganic chlorine sanitisers and it is generally not recommended for use in fluid milk processing plants.

*Iodophors.* In this product, iodine has been combined with non-ionic wetting agents and acidified for stability. Iodophors are generally less corrosive at proper concentrations than chlorine sanitisers.

*Mixed halogens.* Sanitary agents containing both chlorine and bromine are also available. The synergetic action of the two halogens permits lower use levels than those required with regular chemical chlorine sanitisers.

*Quaternary ammonium compounds.* They are non-corrosive to dairy equipment, and their germicidal activity is less affected by the presence of organic matter than other sanitisers. The bacteriocidal effectiveness of quaternary ammonium compounds is influenced by the hardness of the water, and the label should indicate the upper limit of water hardness in which the quaternary sanitiser is effective. They are also less effective against certain spoilage (Gram-negative) bacteria.

*Acid sanitisers.* Acid sanitisers are a mixture of acids and wetting agents, and their germicidal properties are based upon their low pH, and the activity of the wetting agents at this low pH. They are generally slower acting than hypochlorite sanitisers.

*Non-Acceptable Types*

*Phenols and bis-phenols.* Phenol and phenolic compounds have long been known for their antibacterial action. The halogenated bis-phenols are more active than the mono-phenols from which they are derived, e.g.

hexachlorophene, but phenol or phenol derivatives, in general, *are not acceptable types* for use in milk and food processing plants.

*Heavy metals.* Some heavy metals, such as mercury and silver, have a definite germicidal effect on bacteria, but together with the salts of mercury (Hg), silver (Ag), lead (Pb), zinc (Zn), copper (Cu) and chromium (Cr), are *not regarded as acceptable types* for use in milk or food processing plants.

## Special Applications

### *Ultra-violet*

Ultra-violet (UV) radiation has been used with success in the milk and food industry for the reduction of bacteria, fungi and viruses and some practical applications of this approach are: (1) protecting air intakes for laboratory, culture transfer and cultured product processing areas; (2) cleansing the air spaces in liquid sugar tanks; (3) the reduction of air-borne organisms in 'usually occupied' locations and (4) the radiation of packaging material prior to filling, e.g. Tetra-Pak and Pitcher-Pak applications.

Since the action of UV on micro-organisms depends upon the quantity of radiation that reaches the organisms, the system must be properly engineered and maintained. A lighting engineer should be consulted to design the proper spacing, light radiation and protection for employees; exposure to UV radiation can cause severe eye damage.

It must be remembered that UV radiation will not work effectively unless environmental sanitation is maintained.

### *Hydrogen Peroxide ($H_2O_2$)*

Hydrogen peroxide is a strong oxidising agent, but is not considered a strong bacteriocide. It does, however, have the ability to change the environment such that it becomes unsuitable for the growth of organisms, and it has found an application in dairy processing as a headspace mist in Pure-Pak packaging of UHT products. A 15% solution of $H_2O_2$ is recommended, and analytical data should be obtained so that the proper concentration is used; higher concentrations can cause carry-over into the product, and a positive growth inhibitor (GI) test will result. Extreme care should be exercised in handling hydrogen peroxide, as it is a strong oxidising agent and is potentially explosive.

## REFERENCES

AMERICAN PUBLIC HEALTH ASSOCIATION (1975) *Standard Methods for the Examination of Water and Wastewater*, 14th edn. New York.

ANON. (1966) *Pasteurising Plant Manual*, Society of Dairy Technology, London.
BANDLER, D. K. (1979) *What's happening to milk quality?* Cornell University, Ithaca, NY.
BELL, J. A., BECK, M. D. and HEUBNER, R. J. (1950) *J. Am. Medical Assoc.*, **142**, 868.
BLOCK, S. (1968) In: *Disinfection, Sterilisation and Preservation*, Lawrence C. and Block, S. (Eds.), Lea and Febiger, Philadelphia.
BURTON, H. (1965) *J. Soc. Dairy Technol.*, **18**, 58.
HEDLUND, B. (1976) *Nordeuropaeisk Mejeri-Tidsskriff*, **42**, 244.
HSU, D. A. (1970) *Ultra-high temperature processing and aseptic processing*, Damana Tech. Inc., New York, NY.
JAYNES, H. O. (1975) *J. Milk and Fd Technol.*, **38**(7), 356.
LAHR, A. J. (1973) *Dairy cleanser and sanitiser programmes for milking equipment*, Cornell University, Ithaca, NY.
LAWRENCE, C. (Ed.) (1950) *Surface Active Quaternary Ammonium Germicides*, Academic Press Inc., NY.
LEE, A. Y. (1975) *A study into methods of cleaning and sanitising ultrafiltration membranes*, M. S. thesis, Cornell University, Ithaca, NY.
MAXIE, R. (1964) *J. Milk and Fd Technol.*, **27**, 135.
MORTON, E. (1968) In: *Disinfection, Sterilisation and Preservation*, Lawrence, C. and Block, S. (Eds.), Lea and Febiger, Philadelphia.
NILSSON, R. and WILLART, S. (1960) *Milk and Dairy Research*, (*Alnarp*), Report No. 64.
NORTHEAST DAIRY PRACTICES COUNCIL (1972) *Guidelines for the Cleaning and Sanitising of Dairy Farm Equipment*.
O'SULLIVAN, A. (1971) *Dairy Industries Internat.*, **36**(11), 636; **36**(12), 691.
PORTER, M. and MICHAELS, A. (1970) *Proc. 3rd Internat. Congress of Food Science and Technology*, Washington, DC.
RAHN, O. (1945) *Injury and Death of Bacteria by Chemical Agents*, No. 3 of the Biodynamica Monographs, Biodynamica, Normandy, Mo.
RUDOLPH, A. and LEVINE, M. (1941) *Factors affecting the germicidal efficiency of hypochlorite solutions*, Bull. No. 150, Engineering Experiment Station, Iowa State College.
RUEBLE, G. and BREWER, C. (1931) *US food and drug administration methods of testing antiseptics and disinfectants*, USDA Circular 198.
SIEGENTHALER, E. J. (1965) *Von der Eidgenössischen Technischen Hochschule*, Zurich, Prom. Nr. 3695.
SPECK, M. L. and Adams, D. M. (1976) *J. Milk and Fd Technol.*, **37**, 269.
THOMAS, S. B. and THOMAS, B. F. (1973) *Dairy Industries Internat.*, **38**, 61.
US PUBLIC HEALTH SERVICE (1965) *Grade A pasteurised milk ordinances*, US Dept. of Health, Washington, DC.
WESTHOFF, D. L. (1978) *J. Fd Protection*, **41**(2), 122.
WESTHOFF, D. C. and DOORES, S. (1976) *J. Dairy Sci.*, **59**, 1003–9.
ZALL, R. R. (1977) *Environmental Protection Technology Series*, EPA 600/2-77-118.
ZALL, R. R. (1979) *8th Marschall Internat. Dairy Symp.*, London, England.
ZALL, R. R. and BROWN, D. P. (1976) *Am. Soc. Agric. Engineers*, (Winter Meeting), Paper 76-3568, Chicago, Illinois.

ZALL, R. R., GOLDSMITH, G., HORTON, B., HUSSAIN, S. and TAN, M. (1971) *Environmental Protection Technology Series*, EPA Document 12060 DXF, Cincinnati, Ohio.

ZALL, R. R., PRICE, D. R. and BROWN, D. P. (1975) *Proc. 68th Annual Meeting Am. Inst. of Chemical Engineers*, Los Angeles, California.

# 4

# The Microbiology of Raw Milk

Christina M. Cousins and A. J. Bramley
*National Institute for Research in Dairying, Shinfield, UK*

Cows are milked at least twice a day on farms worldwide. The harvesting of milk, a highly perishable foodstuff, varies from the hand milking of a few animals out-of-doors, to the use of large and complex machines for milking herds of 3000 cows in well equipped premises where milking may continue for many hours a day. Under primitive conditions, many small quantities of uncooled milk are taken by producers to a collecting centre but, where dairy farming is more highly developed, an increasingly large proportion of milk is refrigerated immediately after production and stored in tanks on the farm until collection. Thus, the initial microbiological quality of milk is likely to vary enormously. Nevertheless, under any conditions, there are only three main sources of microbial contamination of milk, namely, from within the udder, from the exterior of the teats and udder and from the milking and storage equipment.

Milk is produced at ambient temperatures ranging from sub-zero, where it is necessary to protect milk in cans from freezing to 30 °C or higher where, without refrigeration, it is impossible to cool milk much below 25 °C. Furthermore the temperature and duration of milk storage on the farm can vary widely, so that the numbers and types of micro-organisms present when the milk leaves the farm differ, often unpredictably, even under apparently similar conditions.

During the last 25 years, in most dairying areas, milk production methods, equipment and on-farm storage have changed, generally for the better. However, the microbiological quality of some raw milk supplies, produced under apparently good hygienic conditions and stored under refrigeration, still causes concern because of the possible adverse effects of prolonged refrigerated storage of raw milk and of mastitis, on processed milk and milk products. Refrigeration on the farm all too often masks the effects of unhygienic practices, including the use of inadequately cleaned and disinfected milking equipment; udder disease remains widespread, and consumers of raw milk still risk food poisoning. Factors influencing the microbiological quality of raw milk have been extensively studied by many workers, and have been reviewed in a bulletin issued by the International Dairy Federation (1980).

## THE INITIAL MICROFLORA OF RAW MILK

The numbers and types of micro-organisms in milk immediately after production, i.e. the initial microflora, reflects directly microbial contamination during production. The microflora of the milk when it leaves the farm is determined by the temperature to which it has been cooled, the temperature at which it has been stored, the time elapsing before collection and the initial microflora.

Where milk is cooled to and stored at $\leq 4\,^{\circ}C$, the low temperature will normally prevent bacterial multiplication for at least 24 h, and the microflora is, therefore, similar to that present initially.

The 'total' bacterial count or standard plate count is determined by plating (or equivalent procedures) on plate count agar or yeast extract milk agar followed by aerobic incubation for 2–3 days at 30–32 °C. Micro-organisms failing to form colonies will not, of course, be included. Certain groups may be selectively enumerated, e.g. psychrotrophs by incubating plates for 10 days at 5–7 °C, or thermoduric organisms by laboratory pasteurisation of the milk before plating. Selective or diagnostic media may be used for coliforms, lactic acid bacteria, mastitis pathogens, Gram-negative rods (GNR), lipolytic, proteolytic and caseinolytic types, etc.

### 'Total' Bacterial Content

The initial standard plate count (SPC) may range from $< 1000\ ml^{-1}$, where contamination during production is minimal, to $> 1 \times 10^6\ ml^{-1}$ of milk. The micro-organisms present will be derived from one or any combination of the three main sources of contamination, namely the interior of the udder, its exterior or milking equipment. High initial SPCs in milk, e.g. $> 100\,000\ ml^{-1}$, are evidence of serious faults in production hygiene, whereas the production of milk having SPCs consistently $< 10\,000\ ml^{-1}$ reflects good hygienic practices (International Dairy Federation, 1974).

A widely adopted standard for Grade A or Grade 1 raw milk is an SPC of

$<1 \times 10^5$ ml$^{-1}$, and this may be obligatory for raw milk intended for heat treatment before liquid consumption, but for milk that is to be consumed raw, a more stringent standard is generally required. In some countries, standards adopted may depend on whether milk is refrigerated or merely water-cooled. In North America SPCs of $\leq 3 \times 10^6$ ml$^{-1}$ or equivalent are acceptable for manufacturing grade milk, but in the UK for example, no distinction is made between raw milk going for manufacture and that for liquid consumption. Tests and standards for the hygienic quality of bulk cooled milk are discussed by Lück (1972) and Thomas and Thomas (1957*b*).

A survey of the bacteriological quality of daily collected refrigerated bulk tank milk from *c*. 350 farms (representative of some 18 000 bulk tank milk producers in England and Wales) sampled once a month for a year (Panes *et al*., 1979) has provided one of the most up-to-date studies on the distribution of initial total colony counts (Table I). The geometric mean

TABLE I

INITIAL TOTAL COLONY COUNTS IN BULK TANK MILK SAMPLES FROM 333 FARMS TAKEN AT MONTHLY INTERVALS FOR 12 MONTHS

| *No. of results* | *Percent frequency distribution of cfu ml$^{-1}$ of milk* | | | | | |
|---|---|---|---|---|---|---|
| | *<10 000* | *>10 000–50 000* | *>50 000–100 000* | *>100 000–500 000* | *>500 000–1 000 000* | *>1 000 000* |
| 3 996 | 32·73 | 42·45 | 11·26 | 10·66 | 1·67 | 1·23 |
| | | 86·44 | | | 13·56 | |

Data from Panes (1979).

total colony count for all samples was 19 980 ml$^{-1}$ of milk. Thermoduric, psychrotrophic and pre-incubated colony counts were also determined, together with results of resazurin tests and rinses of milking equipment. Other surveys of bulk milk quality have been reviewed by Thomas *et al.* (1971).

The total bacterial count does not indicate the sources of bacterial contamination in milk, or the identity of production faults leading to high counts. Counts of psychrotrophs, thermoduric organisms, spores, streptococci and coliforms may assist in the diagnosis of faults, but are not infallible. In any case, these additional tests are normally impracticable for routine grading, and they are mainly used for advisory, investigational and survey purposes.

TABLE II

TYPES OF AEROBIC MESOPHILIC MICRO-ORGANISM IN FRESH RAW MILK AND FORMING COLONIES ON MILK COUNT AGARS

| *Micrococci* | *Streptococci* | *Asporogenous Gram +ve rods* | *Spore-formers* | *Gram −ve rods* | *Miscellaneous* |
|---|---|---|---|---|---|
| *Micrococcus* | *Enterococcus* ('faecal') | *Microbacterium* | *Bacillus* (spores or vegetative cells) | *Pseudomonas* | Streptomycetes |
| *Staphylococcus* | | *Corynebacterium* | | *Acinetobacter* | Yeasts |
| | Group N | *Arthrobacter* | | *Flavobacterium* | Moulds |
| | | *Kurthia* | | *Enterobacter* | |
| | Mastitis streptococci | | | *Klebsiella* | |
| | *Str. agalactiae* | | | *Aerobacter* | |
| | *Str. dysgalactiae* | | | *Escherichia* | |
| | *Str. uberis* | | | *Serratia* | |
| | | | | *Alcaligenes* | |

Note: special media and/or incubation conditions are needed for isolation or detection of species of *Clostridium*, *Lactobacillus* and other lactic acid bacteria, *Corynebacterium* and certain pathogens.

**Types of Micro-Organisms Present in Raw Milk**

The main groups of micro-organisms and their components comprising the microflora of milk from individual farms and detected by isolations from plate count agar at 30–32 °C, the aerobic mesophilic flora, are shown in Table II. A relatively simple scheme for characterising isolates, based on morphology, reaction to Gram stain, catalase production and formation of acid and gas in McConkey's broth, first described by Carreira *et al.*, (1955) and used by a number of workers, has provided much useful information on the main groups of micro-organisms in raw milk, as exemplified in Table III for low count raw milk ($<5000 \, cfu \, ml^{-1}$). In such milk, the minimal bacterial contamination from the exterior of the udder and from milking equipment is reflected in the predominance of non-thermoduric micrococci (including staphylococci) and streptococci, and these are presumably bacteria from within the udder.

TABLE III
INCIDENCE OF THE MAIN GROUPS OF MICRO-ORGANISMS IN LOW COUNT RAW MILK

| *Group* | *Incidence* (%) |
|---|---|
| Micrococci | 30–99 |
| Streptococci | 0–50 |
| Asporogenous Gram +ve rods | <10 |
| Gram −ve rods (GNR) (including coliforms) | <10 |
| *Bacillus* spores | <10 |
| Miscellaneous (including streptomycetes) | <10 |

The variations shown in Table III apply not only to low count milk samples from individual farms, but also to mean results of samples from numerous farms (Jackson and Clegg, 1966; Thomas *et al.*, 1962; Thomas, 1974*b*).

As total colony counts increase, then the proportions change; generally, an increase in GNR occurs at the expense of the micrococci, so that the former comprise at least 30% of the microflora. However, in individual samples, micrococci, streptococci or GNR may predominate at any level of total count.

Considerable variation in the incidence of thermoduric organisms and psychrotrophs in fresh raw milk has been reported by different observers. Some differences may be regional or seasonal, and some are associated with methods of cleaning and disinfecting equipment on individual farms. Some variation may be accounted for by differences in methods of performing laboratory pasteurisation for estimation of thermoduric count, and the time and temperature of incubation of plates for both thermoduric and psychrotroph counts (Thomas and Thomas, 1975*a*).

*The Thermoduric Microflora*

The genera surviving laboratory pasteurisation are shown in Table IV. *Microbacterium lacticum* and bacterial spores normally show 100% survival; some *Micrococcus* spp. are slightly less heat resistant, and only 1–10% of strains of *Alcaligenes tolerans* may survive. Species of

TABLE IV

THERMODURIC AND PSYCHROTROPHIC MICRO-ORGANISMS IN FRESH RAW MILK

| *Thermoduric genera*[a] | *Psychrotrophic genera*[b] |
|---|---|
| *Microbacterium* | *Pseudomonas* |
| *Micrococcus* | *Acinetobacter* |
| *Bacillus* spores | *Flavobacterium* |
| *Clostridium* spores | *Aerobacter* |
| *Alcaligenes* | *Alcaligenes* |
| | *Bacillus* |
| | *Arthrobacter* |

[a] Survive heating at 63 °C for 30 min.
[b] Visible growth at 5–7 °C in 7–10 days.

streptococci (e.g. *Str. faecalis*), lactobacilli and some coryneforms are heat resistant, surviving 60 °C for as long as 20 min, but only a small percentage, probably <1%, normally survives 63 °C for 30 min. Most reports of coliforms and strains of *Escherichia coli* surviving pasteurisation are probably attributable to the fact that they have been detected in commercially pasteurised milk. Survival of *E. coli* after heating at 63 °C for 30 min has seldom been substantiated in laboratory tests. The *Bacillus* spore content of raw milk rarely exceeds 5000 $ml^{-1}$ and it is generally higher in winter than in summer, because these organisms are largely derived from surfaces of teats which have been in contact with bedding materials used for housing cows (see section on The Microflora of the Exterior Teats and Udder). Ridgeway (1955) found that 12% of farm milk supplies from winter housed cows had spore counts of >100 $ml^{-1}$, and were much lower in summer. Very heat resistant spores, although only a small proportion of the total spore content, were also more prevalent in winter. In ex-farm raw milk, *B. licheniformis* is the most common species; *B. cereus* is found only sporadically in bulk tank milk. Milk cans are known to be a source of *B. cereus* spores in milk.

In contrast to spores, micrococci and *Microbacterium* spp. are derived almost exclusively from milking equipment, which is sometimes so heavily contaminated with these organisms that thermoduric counts in the milk exceed $5 \times 10^4 \, ml^{-1}$. Most thermoduric organisms do not multiply appreciably in raw milk even at ambient temperatures, and thus a high thermoduric count in milk up to 24 h old is reliable evidence of gross contamination from milking equipment. For this reason, the thermoduric or laboratory pasteurised count has been proposed and, indeed, used for hygienic quality control of raw milk, although the correlation with total count is poor (Lück, 1972). In their survey, Panes *et al.* (1979) found a correlation of 0·65 between thermoduric and total counts when the means for 12 monthly samples from *c.* 350 individual farms were compared; considering individual milk samples, 16·4, 10·9 and 2·7% had thermoduric counts of $>5000$, $>10\,000$ and $>100\,000 \, ml^{-1}$, respectively, but the overall geometric mean thermoduric count was only $750 \, ml^{-1}$ of milk, about 4% of the geometric mean of the initial total counts; there was no evidence of thermoduric counts being higher in summer than in the winter months. Milk delivered in cans had higher thermoduric counts than milk collected from refrigerated bulk tanks (Thomas *et al.*, 1967), but this may have been due to lower levels of thermoduric organisms in the milking machines of the bulk tank producers, rather than to contamination from the milk cans.

Spores of *Clostridium* spp. are thermoduric, but normally can only be detected in raw milk that has been heated to destroy vegetative bacteria followed by the use of suitable media and anaerobic incubation.

Clostridial spore counts are highest in winter, because they are mainly derived from silage used for winter feeding and from bedding materials (see section on The Microflora of the Exterior Teats and Udder). When numbers of spores of *Cl. tyrobutyricum* associated with bad silage are present in milk much in excess of 1 spore $ml^{-1}$, the milk may be unsuitable for making Gruyère and Emmenthal cheese. After cows have gone out to pasture, clostridial spore counts decline, and are usually $<1 \, ml^{-1}$. Clostridia do not multiply in raw milk (Goudkov and Sharpe, 1965).

### *The Psychrotrophic Microflora*

The most commonly occurring psychrotrophs in fresh raw milk are

Gram-negative rods (Table IV). *Pseudomonas* spp. account for about 50% of the Gram-negative genera, and *Ps. fluorescens* predominates; other species include *Ps. putida*, *Ps. fragi* and *Ps. aeruginosa*. *Flavobacterium*, *Acinetobacter*, *Achromobacter*, *Alcaligenes* and coliforms comprise most of the remaining psychrotrophic Gram-negative genera (Juffs, 1973). Some of these psychrotrophs, when growing in refrigerated milk, produce extracellular heat-resistant lipases, as well as proteinases which may degrade casein. Even among strains of one species there is considerable variation in activity; strains of *Ps. fluorescens* are most likely to be both lipolytic and caseinolytic. The Gram-negative psychrotrophs are killed by pasteurisation, but their enzymes are not inactivated.

The incidence of spores of psychrotrophic strains of *Bacillus* spp. in individual producer's milk is low, and seldom exceeds 10 $ml^{-1}$. The species found include *B. coagulans*, *B. circulans*, *B. cereus* and *B. subtilis* (Mikolajcik, 1979). *Arthrobacter* and other Gram-positive species, e.g. streptococci, have also been reported as components of the psychrotrophic microflora, but the latter probably require about 14 days at 7°C to form countable colonies on agar media.

The psychrotrophic microflora derived from teat surfaces is, as yet, a poorly defined, relatively inactive group, but is likely to include spores and coryneform organisms as well as Gram-negative rods. Inadequately cleaned and disinfected milking equipment is the main source of psychrotrophic Gram-negative rods in raw milk. On average they comprise about 10–50% of the initial total count, but in individual samples, the proportion may be much higher. Panes *et al.* (1979) found a correlation coefficient of 0·66 between the psychrotrophic and total milk counts of bulk tank milk samples when mean results from individual farms were compared. In that survey, the geometric mean psychrotroph count, 1305 $ml^{-1}$, was only 7% of the geometric mean total count of *c.* 20 000 $ml^{-1}$. However, 25% of the 5000 samples examined had psychrotroph counts of $>$5000 $ml^{-1}$, and 25% of farms had at least one psychrotroph count of $>$50 000 $ml^{-1}$ during the course of a year. The results of other surveys have been reviewed by Lück (1972) and Mikolajcik (1979). Witter (1961) defined, discussed and reviewed the bacteria multiplying in milk at refrigeration temperatures, which, at that time, were described as psychrophiles.

### *Coliform Organisms*

The incidence of coliforms and *Escherichia coli* in raw milk has received considerable attention, partly on account of their association with contamination of faecal origin, partly because of the spoilage their growth in milk at ambient temperatures can produce, and not least because of the availability of sensitive tests for detecting and enumerating coliforms.

It is now well recognised that the presence of coliforms in raw milk is not evidence of direct faecal contamination, and cannot be relied on to detect failure to clean dirty udders before milking. Coliforms can rapidly build-up in moist, milky residues in milking equipment, which then becomes the major source of contamination of the milk produced. However, relatively low coliform counts in milk do not necessarily indicate effectively cleaned and disinfected equipment. Coliform counts regularly in excess of 100 $ml^{-1}$ are considered by some authorities as evidence of unsatisfactory production hygiene.

Some species of the genera making up the coliform group of bacteria are psychrotrophic (Table IV), and constitute 10–30% of the microflora isolated at 5–7 °C from raw milk; the majority of these coliforms are *Aerobacter* spp. Thomas and Druce (1972) have reviewed and discussed the incidence and significance of coli–aerogenes bacteria in milk.

## UDDER DISEASE AND THE BACTERIAL CONTENT OF RAW MILK

The bacterial content of raw milk may be increased by the presence of mastitis among the producing animals (Cousins, 1978). Mastitis, or udder inflammation, is usually a consequence of bacterial infection, and is responsible for considerable economic loss to the dairy industry because milk yield is reduced (Schalm *et al.*, 1971). The disease may be present in a clinical form, in which macroscopic changes to the milk or the udder are readily detectable by the milker (Fig. 1), but is more commonly present as a sub-clinical condition in which both milk and udder appear normal. Sub-clinical mastitis can only be diagnosed by the examination of milk samples for the presence of pathogenic bacteria, an increased somatic cell count (Fig. 2), or a variety of biochemical changes which are signs of udder disease (Schalm *et al.*, 1971; Wheelock *et al.*, 1966). These compositional changes largely reflect an increased movement of blood components into the milk during inflammation. The milk from infected quarters contains increased concentrations of bovine serum albumin, immunoglobulin, sodium and chloride ions, while concentrations of lactose and potassium are reduced.

Micro-organisms enter the udder through the duct at the teat tip. The duct varies in length from 5 to 14 mm, and its surface is heavily keratinised;

FIG. 1. Withdrawal of foremilk from an udder quarter affected by clinical mastitis, showing severe clotting of the secretion.

milk residues in the form of fat globules and casein micelles are commonly found within it. Certain species of bacteria, most notably *Staphylococcus aureus* readily colonise the teat duct, particularly in the region of the teat orifice. Teat orifice colonisation may persist for many weeks without the bacteria penetrating to the teat sinus and producing mastitis. The mechanisms of penetration of the teat duct by micro-organisms remain a subject for much research, but it has been shown that, under certain conditions, the action of the milking machine can be responsible for propelling bacteria through the teat duct (Thiel *et al.*, 1973). However, since mastitis occurs among beef cattle and unmilked dairy cows, other mechanisms, perhaps involving growth of the bacteria through the duct occur. A survey of approximately 500 British herds in 1977 revealed that

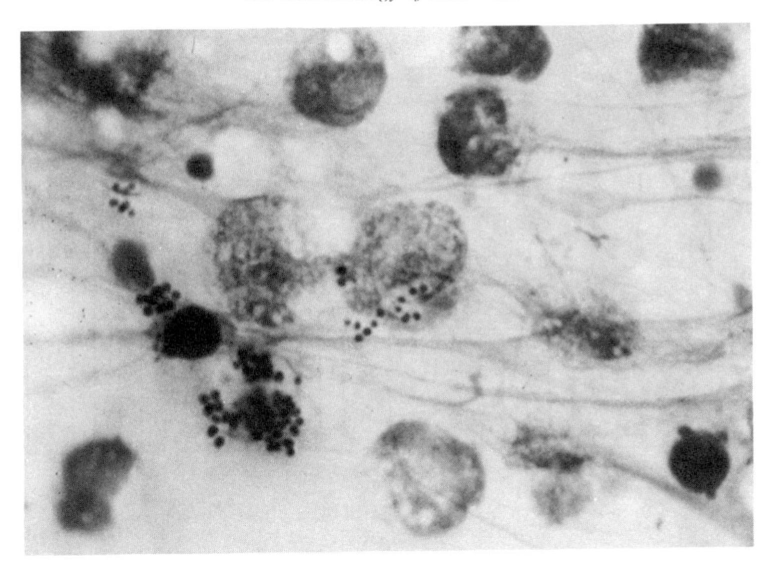

FIG. 2. Microscopic appearance of milk from an udder quarter subclinically infected with *Staphylococcus aureus*, showing infecting cocci, fibrin and increased leucocyte count.

one-third of dairy cattle were infected (Anon., 1979). Olsen reported 12·4 % of cows and 8·1 % of udder quarters infected for a region of Denmark in 1973 (Olsen, 1975). Thirty percent of quarters were found infected among hand-milked cows in Rhodesia (Titterton and Oliver, 1979). However, levels of udder disease vary not only between countries, but also differ considerably between herds (Fig. 3).

Most of the economic loss due to mastitis is a consequence of infection with *Staphylococcus aureus*, *Streptococcus agalactiae*, *Str. uberis*, *Str. dysgalactiae* or *Escherichia coli*. Additionally, *Corynebacterium bovis* and coagulase-negative micrococci are commonly present in milk samples collected aseptically, but rarely produce clinical mastitis or markedly reduce milk yield (Bramley, 1975).

*Streptococcus agalactiae* and *Staph. aureus* are most commonly implicated in mastitis, and are spread between udder quarters and cows primarily during milking, since the major source of the organisms within the herd is the infected udder. The infecting bacteria are excreted in milk and, consequently, the milking clusters, the milkers' hands, udder cloths, etc., become contaminated and may act as fomites transferring disease among the herd. The significance of the diseased udder as the source of the organism has been demonstrated very clearly for *Str. agalactiae* which can

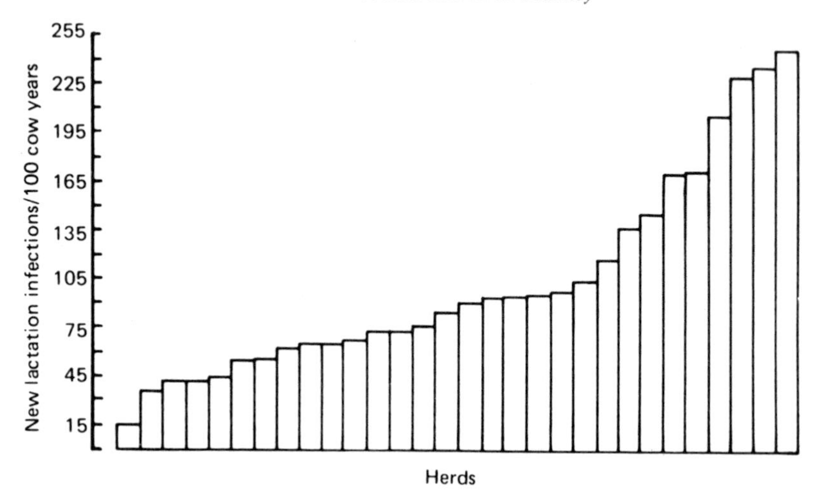


FIG. 3. The variation between herds in the rates of new udder infection during a 3-year period (From: Kingwill *et al.*, 1979.)

be indefinitely eradicated from a herd following the elimination of udder infections. There are data to suggest the same may be true for *Staph. aureus*, although a variety of extramammary sources of this organism exist on the cow, including the vagina and tonsils. (Both *Str. agalactiae* and *Staph. aureus* may be isolated from man and may be transferred to cattle, but only rarely do these human infections act as significant sources for intramammary disease—see the section on pathogens for man in raw milk.)

Other mastitis pathogens differ by being less dependent on the milking process for their dissemination within the herd. Most significant among these are *Str. uberis* and *E. coli* which are widely distributed on the cow or in the environment. *Streptococcus uberis* has been isolated from many sites on the body of the cow, including the lips, teats, rumen, belly, vulva and rectum, and may also be found in large numbers in bedding (Bramley *et al.*, 1979). Bedding and faeces serve as important sources of *E. coli* contamination in outbreaks of mastitis (Bramley and Neave, 1975).

Many countries employ control systems to reduce the incidence of bovine mastitis, and these serve not only to increase the efficiency of milk production, but also to improve the quality of the product which is particularly important to the milk processor. The mastitis control systems used can be broadly divided into two types. In the United Kingdom, United States, Australia and several other countries, control measures are based upon the application of milking time hygiene (most particularly, the application of a disinfectant teat dip after milking), and the treatment of all udder quarters with antibiotic at drying-off. Milking time hygiene and post-milking teat disinfection act to reduce the transfer of bacteria during milking, and to destroy pathogens left on the teat skin at the end of milking. The intramammary infusion of antibiotic at drying-off acts both prophylactically, to prevent new udder disease arising when the cow is not milked, and therapeutically to eliminate existing sub-clinical mastitis. Levels of udder disease have been markedly reduced with this system (Fig. 4), and *Str. agalactiae* eradicated from many herds (Wilson and Kingwill, 1975).

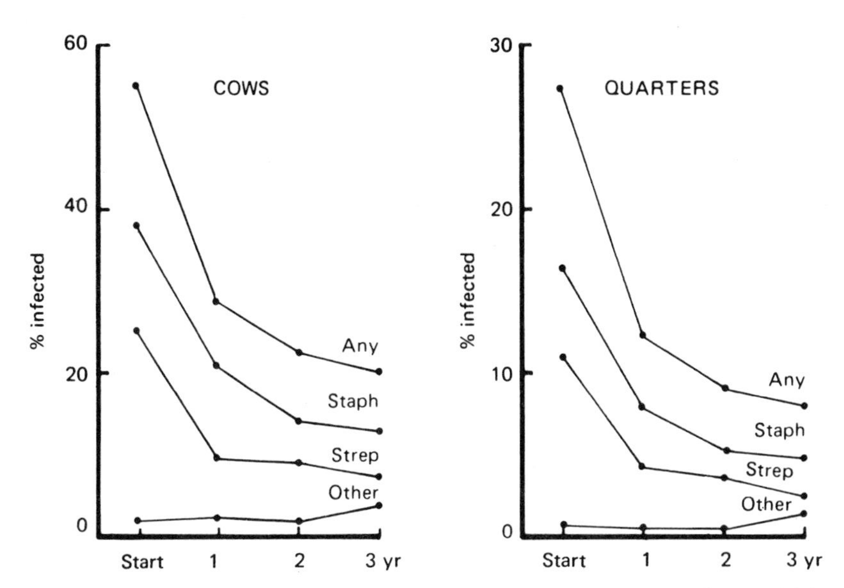


FIG. 4. Reduction in the proportion of cows and udder quarters with sub-clinical mastitis in 30 herds over 3 years using a control system comprising hygiene, including post-milking teat dipping, and antibiotic therapy for cases of clinical mastitis and for all cows at the end of each lactation. (From: Kingwill *et al.*, 1979.)

An alternative approach used in Denmark, Sweden, Switzerland, West Germany and several other countries, employs extensive laboratory facilities to detect heavily infected herds by bacteriological or cytological examination of herd bulk milk. Problem herds are then examined in detail and infected quarters diagnosed and treated with antibiotic (Olsen, 1975). Both approaches have been shown to be effective at reducing rates of streptococcal and staphylococcal udder disease, but are relatively ineffective in preventing *E. coli* mastitis which many veterinarians consider to be an increasing problem, particularly among high producing housed cows.

New udder infections arise most frequently in early lactation, and in the period immediately following drying-off at the end of lactation. Certain pathogens, notably *E. coli* and *Corynebacterium pyogenes*, also show marked seasonal trends. In the case of *E. coli* mastitis, this is due to an association with winter housing (Bramley and Neave, 1975). In Northern Europe, *C. pyogenes* is most commonly encountered as one of the causative organisms of 'summer mastitis', a serious form of the disease prevalent among unmilked cows and heifers between July and September, and probably spread by the head fly, *Hydroteae irritans* (Tarry, 1979). Infected quarters excrete fluctuating numbers of organisms in their milk, and generally the highest numbers of bacteria are found in foremilk (Murphy, 1943). While sub-clinical mastitis usually contributes <10 000 bacteria $ml^{-1}$ of herd raw milk, the inclusion of milk from clinical cases, or quarters shortly to become so, can change this dramatically. Clinical quarters may be excreting >10 000 000 bacteria $ml^{-1}$ and can, under certain circumstances, increase the bulk milk count by 100 000 organisms $ml^{-1}$ (Table V).

TABLE V
THE EFFECT OF INCLUDING MILK FROM A COW INFECTED WITH (*a*) STREPTOCOCCAL, AND (*b*) COLIFORM MASTITIS ON THE BACTERIAL COUNT OF BULK HERD MILK

| | *Count* $ml^{-3}$ *of milk* | | | | | |
|---|---|---|---|---|---|---|
| | (*a*) | | | (*b*) | | |
| *Week sample taken* | *Total bacteria* | *Streptococci* | *Somatic cells* | *Total bacteria* | *Coliforms* | *Somatic cells* |
| 1 | 6·3 | 0·6 | 174 | 29 | 0·01 | 213 |
| 2 | 92·0 | 90·0 | 134 | 200 | 170·0 | 208 |
| 3 | 4·1 | 0·5 | 318 | 35 | 0·02 | 192 |

Data from Cousins, 1978.

We have found this to occur most commonly with cases of streptococcal and coliform mastitis, but only infrequently with staphylococcal udder disease where the numbers of bacteria excreted from infected quarters are generally lower (Table VI).

Mastitis pathogens survive well in raw milk cooled to 10 °C or less, although, with the exception of *Str. uberis*, they will not grow at this temperature. Most will grow at 15 °C in raw milk, although less rapidly than saprophytic contaminants. Bacteriological media commonly used for the

TABLE VI

NUMBERS OF *Staphylococcus aureus* $ml^{-1}$ OF MILK FROM COWS INFECTED IN A SINGLE UDDER QUARTER

| *Cow* | *cfu ml⁻¹ milk* | *Clinical signs of mastitis* | *Cow* | *cfu ml⁻¹ milk* | *Clinical signs of mastitis* |
|---|---|---|---|---|---|
| 1 | 4400 | – | 9 | 19 500 | – |
| 2 | 8 200 | – | 10 | 14400 | – |
| 3 | 1 000 | – | 11 | 26 800 | – |
| 4 | 4 200 | – | 12 | 7 500 | – |
| 5 | 17 300 | + | 13 | 1 300 | – |
| 6 | 9 100 | – | 14 | 8 900 | – |
| 7 | 5 100 | + | 15 | 2 000 | – |
| 8 | 300 | + | 16 | 2 100 | – |

Geometric mean—8730 cfu $ml^{-1}$. Range—300–26 800 cfu $ml^{-1}$.

examination of raw milk will support the growth of the mastitis pathogens, exceptions being *C. bovis* which requires the presence of Tween 80, oleic acid or milk fat in the media, and *C. pyogenes* which needs blood or serum and is stimulated by an atmosphere containing 10% carbon dioxide.

## PATHOGENS FOR MAN IN RAW MILK

Raw milk may contain micro-organisms pathogenic for man, and their source may lie either within or outside the udder. The most important and serious human diseases disseminated by the consumption of contaminated raw milk are tuberculosis and brucellosis. In both diseases the causative organisms, *Mycobacterium bovis* or *M. tuberculosis* (Weir and Barbour, 1950), and *Brucella abortus*, *B. melitensis* or *B. suis* (Smith, 1934), may be excreted in the milk from infected animals. Often with *Brucella* infections, there is little change to the milk or udder, i.e. mastitis is not present. While asymptomatic excretion of tubercle bacilli in milk can occur from infected animals, tuberculous mastitis shows pronounced and characteristic changes to the milk and the udder. In many parts of the world, these diseases have been eradicated from cattle and no longer pose a hazard to human health. Where this has not been achieved, pasteurisation should be employed to destroy these dangerous pathogens and render the milk safe for consumption.

Pathogenic bacteria may also be present in raw milk as a direct consequence of udder disease. Among the organisms commonly producing mastitis *Str. agalactiae*, *Staph. aureus* and *E. coli* are pathogenic for man. *Streptococcus agalactiae* is responsible for a variety of clinical conditions, of which the most serious is an often fatal bacteraemia and meningitis of the newborn. However, the pathogenicity for man of bovine strains of *Str. agalactiae* is uncertain, and the organism is carried by a large proportion of the human population (Sinell, 1973). While it seems unlikely that consumption of contaminated raw milk plays a significant part in infection of the population at large, some researchers have reported higher rates of carriage among consumers of raw milk (Hahn *et al.*, 1970).

Staphylococcal mastitis of the cow poses a more direct threat to public health, because a proportion of bovine strains produce enterotoxin (Olsen *et al.*, 1970). Consumption of food containing enterotoxin leads to an unpleasant illness, usually of approximately 24-h duration, characterised by nausea, diarrhoea and abdominal pain. Although enterotoxin is not thought to be elaborated within the mammary gland, it may be produced if the storage conditions of the milk allow multiplication of the staphylococci. Since enterotoxin is relatively heat stable (Read and Bradshaw, 1966), subsequent pasteurisation of the toxin contaminated milk will not make it safe for consumption.

High numbers of *E. coli* may be present in milk as a consequence of mastitis, and this species is responsible for several different diseases of man of varying severity. While direct links between *E. coli* udder infection and human disease have not been reported, a wide range of *E. coli* serotypes have been isolated from bovine milk, and it is probable that some of these are pathogenic for man.

Infrequently micro-organisms of greater pathogenicity for man produce bovine mastitis and may be present in raw milk. They include *Leptospira* spp., *Listeria monocytogenes*, *Bacillus cereus*, *Pasteurella multocida*, *Clostridium perfringens*, *Nocardia* spp., *Cryptococcus neoformans* and *Actinomyces* spp. Additionally, *Coxiella burnetii*, the causative agent of Q fever, may infect the udder, probably by the haemotogenous route, and contact with, or consumption of, infected milk can lead to human infection.

Further hazards stem from the adventitious contamination of raw milk by pathogenic bacteria from sources external to the udder. Salmonellae and thermoduric *Campylobacter* strains fall into this category, and have produced many outbreaks of enteritis (Robinson *et al.*, 1979; Taylor *et al.*, 1979; Werner *et al.*, 1979). The majority of these stem either directly or indirectly from faecal contamination of the milk. However, it has been demonstrated experimentally that the udder may be infected with

*Salmonella typhi* (Scott and Minett, 1947) or *Campylobacter coli/jejuni* (Lander and Gill, 1979). Human carriers may also be sources of infection in milk-borne outbreaks; this has been reported for *Salmonella* infections, and for cases of scarlet fever or septic sore throat due to *Str. pyogenes* (Bryan, 1969).

All of these pathogens, with the exception of *Cl. perfringens* and *B. cereus* are destroyed by pasteurisation; these two organisms can survive the pasteurisation process because of their ability to sporulate. It is improbable, however, that *Cl. perfringens* can germinate and multiply under the conditions of milk storage and, although *B. cereus* is well recognised as a cause of food poisoning in man, the drinking of contaminated pasteurised milk does not figure as an important cause of *B. cereus* food poisoning; most outbreaks in the UK being associated with the consumption of cooked rice. This restriction may be because the germination of *B. cereus* spores and its growth in pasteurised milk leads to the development of 'off-flavours', and an appearance discouraging consumption (Davies and Wilkinson, 1973); the predominance of non-pathogenic serotypes in milk is also pertinent.

## ANTIMICROBIAL SYSTEMS IN MILK

Several antimicrobial systems are detectable in milk operating either for the protection of the mammary gland from infection, or conferment of disease resistance on the suckling young.

### Immunoglobulin (Ig)

Antibodies to potentially pathogenic bacteria are often present in milk, and they may be produced locally within the udder (IgA), or be transferred to milk from the circulation (IgG). The primary function of these antibodies is protection of the newborn by passive transfer, but the complement/antibody system does operate within the udder to protect it from infection by strains of coliform bacteria which are susceptible to complement/antibody killing (Carroll *et al.*, 1973). Antibodies may also serve to reduce the severity of udder disease, e.g. by neutralising toxins elaborated during the disease process, or by acting as opsonins to facilitate the phagocytosis of bacteria by polymorphonuclear leucocytes (PMN).

### Phagocytosis

It is generally accepted that protection of the udder from mastitis rests primarily on phagocytosis, and the killing of invading bacteria by PMN.

The total cell count of milk from uninfected udders ranges from *c*. 100 000 to 500 000 cells $ml^{-1}$, of which approximately 10 % are PMN. Infected quarters may excrete milk containing 10 000 000 cells $ml^{-1}$, of which 90 % are PMN. The increase in the cell content of milk associated with mastitis allows the assessment of herd levels of udder disease by measurement of the cell content of the herd bulk milk (Westgarth, 1975). Phagocytosis and killing by PMN is less effective in milk than in blood, largely because the PMN ingest large quantities of fat and casein (Russell *et al.*, 1977). As a consequence of this reduced phagocytic activity, the udder is relatively easily infected even by small numbers of invading bacteria. Increasing the PMN content of milk has been shown to increase resistance of the udder to infection (Schalm *et al.*, 1964). Additionally, depleting the cow of PMN by the use of equine antibovine PMN serum, resulted in chronic sub-clinical staphylococcal mastitis being converted to a peracute gangrenous form of the disease (Schalm *et al.*, 1976).

**Non-Specific Defence Mechanisms**

Several other antibacterial systems, which have a broad spectrum of activity, also occur in milk.

*Lactoferrin* (*LF*)

LF is an iron-binding protein similar to serum transferrin, it is present, and its concentration is markedly increased in the secretion from unmilked or infected animals.

Lactoferrin inhibits the multiplication of bacteria by depriving them of iron, and may protect the dry udder from infection with *E. coli* (Reiter and Bramley, 1975). Although LF is present in bovine milk, the high citrate and low bicarbonate concentrations have been shown to markedly reduce the iron-binding and, therefore, the inhibitory properties of LF.

*Lactoperoxidase/Thiocyanate/Hydrogen Peroxide System* (*LP System*)

Lactoperoxidase is synthesised within the mammary gland, and is present in high concentration in bovine milk. Thiocyanate will be present in varying concentrations related primarily to the nutrition of the cow. The third component of the system, hydrogen peroxide, may be supplied by hydrogen peroxide-producing organisms within the udder (e.g. streptococci), or by the PMN. The LP system will only temporarily inhibit the growth of some organisms (e.g. group B and N streptococci), while having bacteriocidal activity for others (e.g. group A streptococci, *E. coli* and *Salmonella typhimurium*). It is doubtful whether the LP system plays a role in the defence of the mammary gland, although it may contribute to the protection of the calf from enteritis. It has been suggested that the LP system might be utilised as a 'cold sterilisation' process to render milk safe for consumption, without damaging, by heat treatment, the various antimicrobial systems present (Reiter, 1978).

Other antimicrobial systems have also been demonstrated in milk including lysozyme (Vakil *et al.*, 1969) and vitamin binders for $B_{12}$ and folate (Ford, 1974). The significance of these systems for the protection of the mammary gland or the neonate remain unknown.

## THE MICROFLORA OF THE EXTERIOR UDDER AND TEATS

Between milkings, cows' teats may become soiled with dung, mud and bedding materials, such as straw, sawdust, wood shavings or sand. If not removed beforehand this dirt on the teats together with the large number of micro-organisms associated with it, is washed into the milk during milking. Numbers and types of micro-organisms vary according to the type and amount of soil on the teats, and milk from cows with unwashed teats heavily soiled with dung may have a bacterial count approaching 100 000 cfu $ml^{-1}$.

### Numbers of Micro-Organisms from Teats

Bedding materials on which cows are housed in winter may have very high bacterial counts, $10^9$–$10^{10}$ cfu $g^{-1}$, although the bedding may appear relatively clean and dry. The incidence of the main groups of micro-organisms in bedding is shown in Table VII (Cousins, 1978). Particles of these heavily contaminated materials adhere, sometimes unobtrusively, to teat surfaces, and bacterial counts of teat apex swabs of cows kept on sand

TABLE VII

INCIDENCE OF DIFFERENT GROUPS OF BACTERIA IN WOOD SHAVINGS, STRAW AND SAND BEDDING

| *Bedding* | *Geometric mean*[a] *(cfu g⁻¹)* | | | |
|---|---|---|---|---|
| | *Total* | *Psychrotrophs* | *Coliforms* | Bacillus *spores* |
| Shavings | $1{\cdot}2 \times 10^{10}$ | $1{\cdot}1 \times 10^9$ | $8{\cdot}3 \times 10^5$ | $5{\cdot}4 \times 10^6$ |
| Straw | $7{\cdot}4 \times 10^8$ | $9{\cdot}8 \times 10^7$ | $1{\cdot}8 \times 10^5$ | $1{\cdot}5 \times 10^5$ |
| Sand | $5{\cdot}4 \times 10^9$ | $1{\cdot}4 \times 10^9$ | $3{\cdot}9 \times 10^5$ | $5{\cdot}0 \times 10^6$ |

[a] Six samples of each type of bedding.

show the extent to which apparently clean, unwashed teats may be contaminated (Table VIII).

Teats of cows kept in strawed yards can become heavily and visibly soiled if the straw is not adequately and frequently replenished, and bacterial contamination of the teats is correspondingly high unless teats are thoroughly washed.

TABLE VIII

EFFECT OF HOSE WASHING AND DRYING OF TEATS ON BACTERIAL COUNTS OF TEAT APEX SWABS OF COWS BEDDED ON SAND OR ON PASTURE

| *Conditions* | *Herd* | *Teats* | *Geometric mean*[a] *(cfu per teat apex)* | | | |
|---|---|---|---|---|---|---|
| | | | *Total* | *Psychrotrophs* | *Coliforms* | *Spores* |
| Bedded on sand | A | Unwashed | $8 \cdot 4 \times 10^6$ | $1 \cdot 2 \times 10^6$ | 10 | $5 \cdot 0 \times 10^4$ |
| | | Washed | $7 \cdot 3 \times 10^5$* | $8 \cdot 3 \times 10^4$ | 12 | $1 \cdot 2 \times 10^4$ |
| | B | Unwashed | $3 \cdot 3 \times 10^7$ | $1 \cdot 3 \times 10^6$ | 15 | $1 \cdot 0 \times 10^5$ |
| | | Washed | $8 \cdot 5 \times 10^6$* | $4 \cdot 0 \times 10^5$ | 11 | $4 \cdot 8 \times 10^4$ |
| On pasture | A | Unwashed | $7 \cdot 5 \times 10^4$ | $1 \cdot 2 \times 10^4$ | 1 | $1 \cdot 3 \times 10^2$ |
| | | Washed | $3 \cdot 1 \times 10^4$ | $2 \cdot 5 \times 10^3$ | 9 | $1 \cdot 1 \times 10^2$ |
| | B | Unwashed | $1 \cdot 2 \times 10^5$ | $4 \cdot 3 \times 10^3$ | 14 | $4 \cdot 9 \times 10^2$ |
| | | Washed | $1 \cdot 4 \times 10^5$ | $3 \cdot 3 \times 10^3$ | 11 | $5 \cdot 8 \times 10^2$ |

[a] Mean of six samples
* $P < 0 \cdot 05$

In summer, when cows are turned out to pasture, a marked decline in the level of contamination on teats occurs (Table VIII). This seasonal effect is reflected in a marked decline in the bacterial count of the bulk herd milk in summer, especially on farms where milking equipment is effectively cleaned and disinfected so that bacterial contamination from this source is minimal. Bacterial counts of bulk tank milk from a herd at NIRD (National Institute for Research in Dairying) showed that even where cows teats were regularly washed in both winter and summer, winter housing increased the numbers of cfu per ml of the milk (Table IX).

The numbers of bacteria remaining on teat surfaces after washing can be remarkably high, as demonstrated by swabs or rinses of teats (Thomas *et al.*, 1971). The effects of hose washing, using a solution of hypochlorite, *c.* 600 ppm available chlorine, followed by drying teats with paper towels, in reducing teat swab counts where cows were bedded on sand in cubicles are shown in Table VIII, but even after washing, total counts of $10^6$ cfu per teat end were recovered. Reductions for other groups of micro-organisms were

TABLE IX
INFLUENCE OF HOUSING AND TEAT WASHING ON BACTERIAL CONTENT OF BULK TANK MILK FROM ONE HERD SAMPLED ONCE WEEKLY

| *Conditions* | *Teats* | *Geometric mean*[a] *(cfu ml*$^{-1}$ *of milk)* | | | | |
|---|---|---|---|---|---|---|
| | | *Total* | *Psychrotrophs* | *Coliforms* | *Thermoduric organisms* | Bacillus *spores* |
| Bedded on sand | Unwashed | 31 700 | 1 500 | 43 | 120 | 18 |
| | Washed | 15 500 | 990 | 61 | 110 | 14 |
| On pasture | Unwashed | 4 250 | 280 | 19 | 990 | 7 |
| | Washed | 3 530 | 270 | 26 | 750 | 5 |

[a] Each result is the mean of 8–9 milk samples.

not as great, and coliform counts were not affected by washing; these counts were all low, averaging only about 10 cfu per teat (Cousins, 1978). Teat washing using an iodophor solution was ineffective in reducing the bacterial population of the teat apex, unless that procedure was done carefully and was followed by thorough drying (Zarkower and Scheuchenzuber, 1977). Differences in effectiveness between various types of disinfectant solutions used for teat washing in reducing bacterial contamination on teat ends have not been clearly demonstrated, probably because of the over-riding effects of drying after washing.

The direct effects of teat washing on bacterial counts in milk have to be studied using only cows free from udder infections (i.e. giving milk containing $<10$ cfu ml$^{-1}$ in aseptically drawn samples) and, also, milked with sterilised milking equipment or in-line milk samplers. Where these precautions were taken, average total counts ranged from 7000 cfu ml$^{-1}$ in milk drawn from teats unwashed or washed with water and left wet, to 1500 cfu ml$^{-1}$ in milk from teats hose-washed rapidly with hypochlorite solution and dried with paper towels (Cousins, 1978). These average differences concealed widely varying counts in milk from individual cows (Table X). This is because some cows' teats were, before washing, sometimes relatively clean even when housed on very dirty bedding. Conversely, under clean conditions, individual teats were occasionally heavily soiled with dung. Milk from washed teats also had variable counts, probably because it was difficult to ensure that teat ends were clean in the limited time (15 s) available for washing. The difficulty of controlling both environmental conditions and effectiveness of teat washing probably explains why, in various surveys ratings of 'efficiency', obtained by inspection of teat

TABLE X

CONTAMINATION OF MILK WITH BACTERIA FROM THE SURFACES OF COWS' TEATS, EITHER UNWASHED OR AFTER HOSE-WASHING

| *Treatment of teats* | *cfu ml$^{-1}$ of milk (geometric mean counts of 30 samples)* | | |
|---|---|---|---|
| | *Total count* ($\times 10^3$) | *Spores* | *Coliforms* |
| Unwashed | 7·5 (0·5–75·6) | 34 (4–555) | 2 (0–20) |
| Washed with water, left wet | 7·9 (0·6–111·0) | 31 (3–590) | 1·3 (0–10) |
| Washed with water, dried | 4·2 (0·1–54·0) | 16 (1–137) | 0·5 (0–4) |
| Washed with NaOCl, left wet | 4·1 (0·4–64·2) | 38 (6–180) | 0·7 (0·4) |
| Washed with NaOCl, dried | 1·5* (0·1–22·0) | 14 (2–112) | 0·03 (0–1) |

In brackets: ranges of counts.
* Significantly different, $0{\cdot}01 < P < 0{\cdot}05$.

washing and udder and teat cleanliness, have generally failed to show a clear relationship with bacterial counts in the milk (Lück, 1972; Panes *et al.*, 1979). Where only one herd was studied, however, samples of the bulk herd milk from cows bedded on sand had much higher total bacterial counts when teat washing was omitted, as compared with counts of samples taken during the normal teat washing procedure (Table IX). During the summer when cows were on pasture, omitting teat washing had little effect on total bacterial counts which were much lower than in winter (Table IX).

**Types of Micro-Organisms from Teat Surfaces**

In winter, for housed cows, total bacterial counts range from $10^5$–$10^7$ cfu per teat; micrococci including coagulase-negative staphylococci are among the predominent groups present, *c.* $10^4$ cfu per teat. Streptococci, mainly faecal types, are also numerous but Gram-negative bacteria, including coliforms, are much less numerous, and coliform counts rarely exceed $10^2$ cfu per teat (Thomas *et al.*, 1971; Cousins, 1978). It appears that these organisms, unlike micrococci, for example, do not survive well on teat skin, although they form a large proportion of the microflora of bedding materials. Psychrotrophs (detected by incubation of plates at 5 °C for 10 days) range from $10^3$ cfu per teat for washed teats of cows at pasture, to $10^6$ cfu per teat for unwashed teats of cows bedded on sand (Table VIII).

The psychrotrophic microflora of teat surfaces is a poorly defined group of organisms, consisting of coryneforms and Gram-negative rods, most of which are inactive in litmus milk at 22 °C, and which do not appear to multiply readily in raw milk (Johns, 1962; 1971).

The aerobic thermoduric organisms on teat surfaces are almost entirely *Bacillus* spores, spore counts ranging from $10^2$–$10^5$ per teat depending on environmental conditions. The predominant species derived from this source are *B. licheniformis*, *B. subtilis* and *B. pumilis; B. cereus*, *B. firmus* and *B. circulans* occur less frequently (Underwood *et al.*, 1974). Table X shows that aerobic spore counts in milk from individual cows ranged from 1–590 ml$^{-1}$, but spore counts up to 3000 ml$^{-1}$ have been reported (Underwood *et al.*, 1974) in milk from commercial farms.

It can be deduced that teat surfaces are also a source of clostridial spores in milk. They have been detected in cows' fodder, bedding and faeces, and decline markedly in numbers when cows go out to pasture. Spores of lactate fermenting clostridia (*Cl. tyrobutyricum*), which may cause faults in Dutch, Emmenthal and Gruyère cheese, are derived from 'bad' silage and may be transmitted via faeces of silage fed cows to the cows' teats and thence to the milk unless the faecal material is washed from the teats (Bergère, 1979).

A wide variety of genera and species of micro-organisms in the cow's environment may be present on teat surfaces, but those forming only a small proportion of the microflora will not be detected. They will, however, be transmitted to milk and hence to the milking equipment. Some may become established there if conditions are suitable; others may be only transient contaminants. Normally only genera and species capable of forming colonies on plate count agar incubated aerobically are detected. Specific cultural conditions are needed to permit growth of, for example, lactobacilli, clostridia, and other fastidious micro-organisms from dung, soil, herbage and water. Fresh raw milk may be inhibitory or even bacteriocidal to some micro-organisms derived from teat surfaces, so that even if present in appreciable numbers, they may not be detected. A close relationship between the proportions of different types on teats and those present in milk is not, therefore, to be expected.

Removing dirt by filtration of milk, a commonly recommended practice, will not remove particles the size of fat globules or smaller, or somatic cells. Therefore, bacteria introduced into the milk pass through the filter and remain in the milk. Hence, bacterial contamination of filtered milk is not closely related to sediment test gradings.

Essential measures to minimise bacterial contamination from teat surfaces are: action to prevent cows' teats from regularly becoming heavily soiled between milkings, and washing and drying teats before milking.

## AERIAL CONTAMINATION

Air is not an important source of micro-organisms in milk, although small numbers may fall into the milking bucket during hand milking, or they may be carried into milk by air entering the milking machine during use.

Bacterial counts of air in cowsheds or parlours seldom exceed 200 cfu litre$^{-1}$, and are usually much less. Micrococci account for $>50\%$ of the aerial microflora, coryneforms, *Bacillus* spores and small proportions of streptococci and Gram-negative rods will also be present. Calculations based on recorded levels of aeriel contamination, and on the volume of air passing into the machine, indicate that, normally, airborne bacteria would account for $<5$ cfu ml$^{-1}$ of milk produced, and that *Bacillus* spores would constitute $<1$ cfu ml$^{-1}$ (Benham and Egdell, 1970; Underwood *et al.*, 1974). Such levels of contamination are negligible in comparison with those derived from teat surfaces.

## THE MILKER

When cows are hand milked, it is probable that the actions of the milker, in dislodging dust and dirt particles, by increasing aerial contamination in the environment of the udder or by contact with hands, may add micro-organisms to milk. Risks of contamination from the milker are much less with machine milking but, because of the possibility of infection of milk with pathogens, many countries have regulations by which persons known to be suffering from certain diseases are not permitted to take part in the production of milk.

## WATER SUPPLIES

Water used in the process of milk production should be of bacteriologically potable quality. The purity of properly treated supplies direct from the mains is assured, but bacterial contamination can be introduced from storage tanks which are not properly protected from rodents, birds, insects and dust. Bacteria may also come from dirty wash troughs, carrying buckets and hoses.

Many farms rely on untreated water supplies from boreholes, wells, lakes, springs and rivers; some of these may be contaminated at source with micro-organisms of faecal origin, e.g. coliforms, faecal streptococci and clostridia. In addition, a wide variety of saprophytic micro-organisms derived from the soil, or from vegetation may be present, including

*Pseudomonas* spp., coliforms and other Gram-negative rods, *Bacillus* spores, coryneform bacteria and lactic acid bacteria. Numbers of these contaminants will vary widely.

If untreated water gains access to milk or is used for rinsing equipment and containers, any micro-organisms present in the water will contaminate the milk, although the numbers of micro-organisms added, even from relatively heavily contaminated water, may be insignificant in terms of cfu $ml^{-1}$ of milk. However, multiplication of some of the water-borne bacteria in any residual water in the equipment will result in more serious contamination and may lead to the establishment and development of some undesirable types of micro-organism, e.g. psychrotrophic Gram-negative rods, in the milking equipment.

For these reasons, in countries where farm water supplies are known to be bacteriologically unsatisfactory, chemical disinfection or sanitisation of milking equipment is always delayed until just before the next milking, and the disinfectant solution is merely drained from the equipment before it is used for milking. This practice prevents recontamination resulting from rinsing with untreated water.

Chlorination, by dosing with hypochlorite, is frequently recommended for water of unsatisfactory bacteriological quality used for the final rinsing of equipment, because it helps to reduce the risk of bacterial multiplication in residual water left in milking machines that are cleaned and sanitised in one operation.

Warm water for hose washing of udders and teats is often supplied from a water tank controlled at about 37 °C. Some types of bacteria derived from the water entering the tank, or gaining access to an inadequately protected tank, can multiply in the warm water. Udder washing water, thought to have become heavily contaminated in this way with *Pseudomonas* spp. and coliforms, is believed to have been responsible for outbreaks of mastitis caused by these organisms. The risk of such udder infections can be reduced by entraining a disinfectant, e.g. hypochlorite or iodophor into the hose water; this will also help to reduce the numbers of bacteria left on teats after udder washing.

## THE MICROFLORA OF MILKING EQUIPMENT AND ITS EFFECTS ON RAW MILK

Inadequately disinfected (sanitised) milk contact surfaces of milking equipment, including milk cans and bulk tanks, are the only major sources of bacteria in milk after it leaves the udder until collection.

The simplest form of milking equipment is a bucket used for hand milking. Machine milking requires one or more teat cup clusters which are used for milking (a) into buckets or cans, (b) direct into a milk pipeline, or (c) into recorder jars which subsequently discharge the milk into a pipeline (Fig. 5). This pipeline transports the milk to a receiver from which it is released or pumped into milk cans or a bulk tank. Ancillary equipment includes (a) a strainer or in-line filter, (b) a cooler which may be open surface, in-can, or a plate heat exchanger, and (c) milk flow indicators and milk meters (Akam, 1979).

Cows are normally milked twice daily, and the milking machine has to be cleaned after each milking. Because of the complexity of milking machines and some of their components, cleaning and, in particular, disinfection may not be fully effective, so that milk residues and associated bacteria are not completely removed from the equipment and tend to accumulate daily. Except in very cold weather, bacteria in the equipment multiply between milkings, and their numbers may increase more rapidly than visible residues; unfortunately bacterial contamination cannot be determined simply by inspection.

In practice, the contribution of milking equipment to the microflora of the milk cannot be accurately assessed by bacterial counts on the milk produced, because of the variability in numbers and types derived from cows' udders. The most effective method of determining the extent of bacterial contamination on milk contact surfaces of equipment is by rinsing using a sterilised liquid, and then subjecting the rinse to a bacterial count in the same way that milk is tested. The types of bacteria in the rinse can also be determined.

The exact relationship between numbers of bacteria recovered by rinsing a milking machine, for example, and the numbers it contributes to milk during milking is not known. However, from results of repetitive rinsing and by other means (Cousins, 1963; 1972), the proportion recovered by rinsing is known to be at least 10% of the number available to the milk. The proportion will be higher where bacteria have multiplied on moist surfaces or in residual water that has collected in badly drained machines, because this type of contamination is readily rinsed from the surfaces.

It has frequently been pointed out that milking equipment must be very heavily contaminated to increase markedly the bacterial count per ml of the milk passing through it. For example, to increase the bacterial count of 1000 litres (*c.* 200 gal.) of milk by 1 bacterium $ml^{-1}$ requires 1 million bacteria; so, to increase the count by 10 000 $ml^{-1}$, requires 10 000 million bacteria. A milking installation, consisting of four or five teat cup clusters

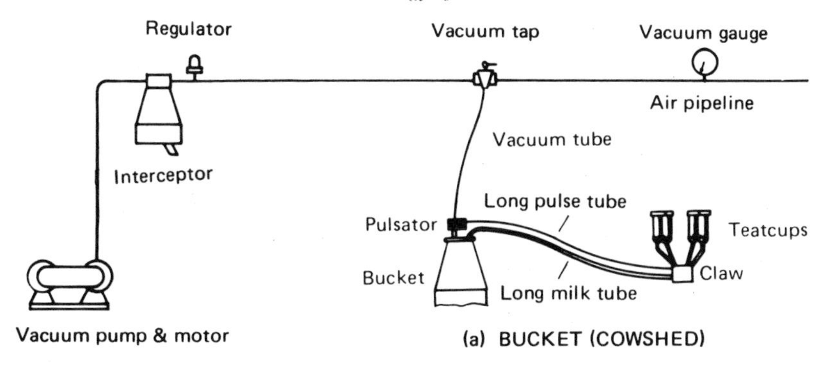


(a) BUCKET (COWSHED)

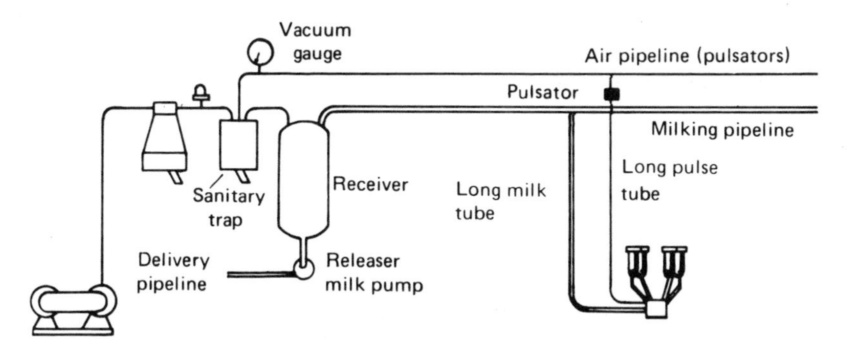


(b) MILKING PIPELINE
(COWSHED & PARLOUR)

Air pipeline (pulsators)
Air pipeline (milking vacuum)
Transfer pipeline
Recorder
jar
Transfer
tube

(c) RECORDER (PARLOUR)

FIG. 5. Diagrams of principal types of milking machines used in the UK. (From: Akam, 1979.)

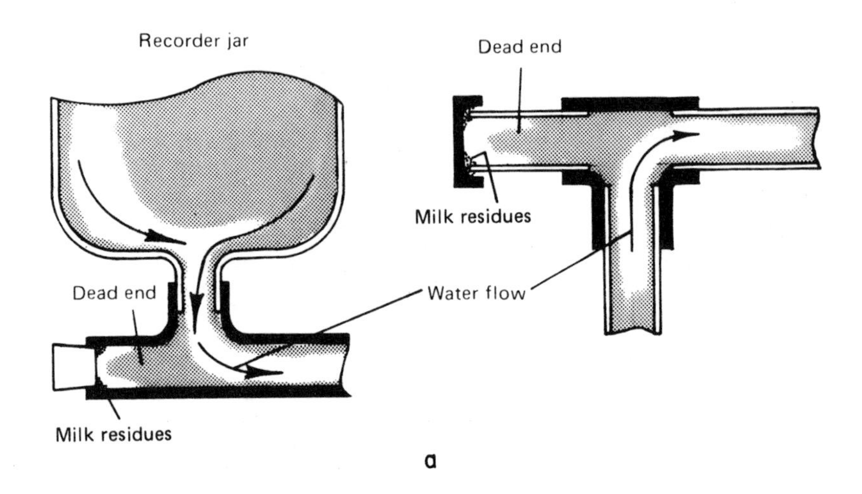


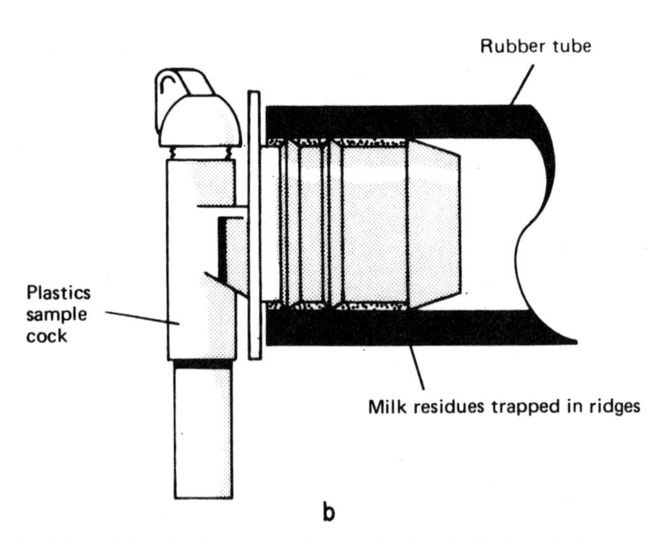


FIG. 6. (a) Milk residues in components forming dead ends: (Left) at the base of a recorder jar, and (right) at the end of a transfer pipeline. (b) Section of a plastic sample cock at the base of a recorder jar showing site of milk residues. (c) Section of a receiver and lid showing sites of milk residues. (From: Cousins and McKinnon, 1979.)

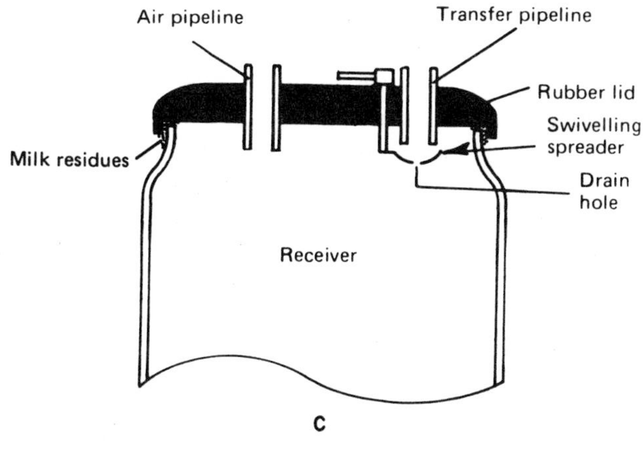


FIG. 6—*contd.*

and jars together with pipelines and a bulk tank, has a milk contact surface area of roughly 10 $m^2$ (*c.* 100 $ft^2$) and, therefore, would need to contribute 1000 million bacteria $m^{-2}$ (or 100 million $ft^{-2}$), on average, of its surface area. Clearly, where milking equipment is solely responsible for high counts, e.g. $>$50 000 cfu $ml^{-1}$, in the milk, cleaning and disinfection must be seriously defective. However, milking equipment is seldom uniformly contaminated; the bacteria and milk residues accumulate in difficult-to-clean areas, and in parts of badly designed components. Figure 6(a), (b) and (c) shows examples of these danger points: crevices, joints, dead ends and fittings which need to be dismantled at intervals because they cannot be effectively cleaned in place. In contrast, the smooth surfaces of jars and pipelines are readily cleaned by circulation of solutions through the machines.

**Numbers of Micro-Organisms on Milking Equipment Surfaces**

In the UK, bacteriological rinses and swabs of milking equipment have been used for many years for advisory, investigational and survey purposes, and the methods are well documented (Cousins, 1963; McKinnon and Cousins, 1969; British Standards Institution, 1975).

Results of rinses taken during the course of advisory work in Wales (Thomas *et al.*, 1966) revealed remarkably high counts in some milking equipment. About 20% of milking machine clusters had rinse counts of $>1 \times 10^9$ cfu per cluster, and one third of the deposits of milky residues scraped from milk tubes contained $>1 \times 10^9$ cfu $g^{-1}$. Many of these high counts were obtained from equipment on farms which had had reports of unsatisfactory milk test results.

Other surveys showed that very good results were attainable where recommended methods of cleaning and disinfection were followed, and that high rinse counts were, in the main, associated with deviations from these methods, the use of rubbers and other equipment in poor physical condition, and the build-up of milk residues.

Milking with bucket machines and the collection of milk in cans is still widespread for small herds, but bulk milk collection from farm bulk tanks, usually refrigerated, and milking by means of pipeline machines have increased rapidly since about 1950.

*Pipeline Milking Machines*

The use of steam sterilisation is now obsolete in the UK because it is costly and time consuming, but it was bacteriologically effective. Soon after reasonably effective systems of in-place cleaning by circulating detergent–disinfectant solutions had been developed, the bacteriological condition of the circulation-cleaned machines was found to compare unfavourably with results obtained using steam sterilisation, when rinse counts of $<5 \times 10^4\ \mathrm{ft}^{-2}$ ($c.\ 5 \times 10^5\ \mathrm{m}^{-2}$) could regularly be achieved. However, the circulation-cleaned machines were generally superficially clean, the new method was labour saving and cheaper than steam, and it quickly gained in popularity. Surveys during the 1960s showed the beneficial effects on rinse counts of circulating very hot detergent–disinfectant solutions (initial temperature 75–80 °C), and this practice is still officially recommended. The temperature of such hot solutions drops by 25–35 °C during circulation, but heat plays some part in the disinfection process in the machine.

Another in-place cleaning system, acidified boiling water (ABW) cleaning, which relies solely on heat for disinfection (Cousins and McKinnon, 1979) is also widely used in the UK. Both these methods, when correctly applied, are capable of effectively cleaning and disinfecting machines so that they contribute relatively few bacteria to the milk passing through them, but it is evident from surveys that some machines are heavily contaminated, probably because of faults in design, incorrect layout of components, incorrect adjustments leading to an unbalanced flow of solutions, or the use of solutions that are not hot enough. Thus numbers of micro-organisms recovered by rinsing these machines can range from $<5 \times 10^5$ to $>1 \times 10^9\ \mathrm{cfu\ m}^{-2}$.

In the course of the survey reported by Panes *et al.* (1979) rinses were taken of the milking equipment (pipeline milking machines and refrigerated bulk tanks) on *c*. 350 farms, once in winter and once in summer. Data extracted from this report show the extent of bacterial contamination in the equipment (Table XI). For milking machines, ABW cleaning gave better results than circulation cleaning but, even so, 4·5 % of rinse counts of ABW cleaned machines were $>1 \times 10^9$ cfu m$^{-2}$. In spite of the fact that 25 % of the machines had what are considered to be very high rinse counts, $>1 \times 10^8$ cfu m$^{-2}$, only 13·5 % of milk samples had total initial counts of $>1 \times 10^5$ cfu m$^{-1}$ (see Table I), and would, therefore, have failed to meet the fairly widely adopted Grade A or Grade 1 hygienic quality standards for ex-farm raw milk.

TABLE XI

BACTERIAL CONTAMINATION IN PIPELINE MILKING MACHINES, BULK MILK TANKS AND ON BULK TANK OUTLET PLUGS

| *Equipment* | *No. of rinses* | *Percentage frequency distribution of cfu m$^{-2}$*[a] | | | | |
|---|---|---|---|---|---|---|
| | | $>1 \times 10^5$ | $>1 \times 10^6$ | $>1 \times 10^7$ | $>1 \times 10^8$ | $>1 \times 10^9$ |
| Milking machines | 702 | 98·3 | 91·0 | 66·7 | 26·1 | 7·4 |
| Tanks, cleaned by: | | | | | | |
| hand (brush) | 284 | 77·7 | 56·2 | 26·5 | 4·9 | 1·1 |
| hand spray | 277 | 80·5 | 50·2 | 23·1 | 4·7 | 0·7 |
| automatic spray | 194 | 56·0 | 33·7 | 14·0 | 2·6 | 0·0 |
| Tank outlet plugs[b] | 755 | 82·5 | 67·9 | 46·7 | 23·8 | 8·9 |

Data from Panes *et al.*, 1979.
[a] $1 \times 10^6$ cfu m$^{-2}$ = 100 cfu cm$^{-2}$ $\simeq 1 \times 10^5$ cfu ft$^{-2}$.
[b] Results expressed as cfu per plug.

About half the machines in the survey were cleaned with hot solutions twice daily, the remainder only once daily, cold solutions being used after evening milking. There was no clear evidence that omitting one of the hot treatments had any detrimental effect on the rinse counts of the machines, and it is probable that a once daily heat treatment, provided it is effectively applied, would prevent any build up of detectable levels of milk residues and bacteria which might result from cold cleaning.

In Ireland, circulation cleaning using a cold caustic-based detergent without a conventional disinfectant is reported to maintain the milk contact surfaces of large pipeline machines in a clean condition for at least a month. The bacteriological results compared favourably with those of conventional circulation cleaning, the incidence of proteolytic Gram-negative rods in the microflora of the plant rinses was low and savings in costs were substantial (Palmer, 1977). The effectiveness of the system seems to be due to the fact that after circulation of the caustic solution, it is drained, but not rinsed, from the machine. This allows the residual solution a long contact time with the milk contact surfaces of the machine, and perhaps facilitates penetration of the caustic solution into joints and crevices. Rinsing the machine, using cold water, is delayed until just before the next milking. A monthly heat treatment is recommended to remove or prevent the formation of deposits.

*Farm Bulk Milk Tanks*

Most farm tanks have smooth, stainless steel surfaces which are more easily cleaned than are milking machines. Accessories such as the agitator, dipstick, plug or outlet cock and, on some types of tank, manhole gaskets can sometimes cause problems. In the UK, most farm tanks of $<4000$ litre (900 gal.) capacity are 'cold wall' or ice bank tanks for which hot solutions cannot be used for cleaning, because they would increase the refrigeration load and, furthermore, would be rapidly cooled on the tank surfaces. These tanks are cleaned by means of mechanical or hand sprays using cold iodophor solutions, or by manual brushing. In North America and some other dairying areas, the tank cooling systems are different from those of UK tanks, and hot cleaning is more usual. However, there is little information on the bacteriological condition of hot cleaned tanks.

Until about 1970, total bacterial counts obtained by rinsing and swabbing tanks and their accessories, whether cleaned manually or mechanically (automatically), were in the main $<2{\cdot}5 \times 10^6$ cfu m$^{-2}$, the majority being $<1 \times 10^5$ cfu m$^{-2}$, and mechanical cleaning was rated rather more effective than manual cleaning (Druce and Thomas, 1972). This observation has been confirmed by Panes *et al.* (1979), although their results indicate a deterioration in the bacteriological cleanliness of farm tanks as compared with earlier results; about 30% of their tanks had rinse counts of $>5 \times 10^6$ cfu m$^{-2}$ (Table XI). On about 1% of farms, the tanks were grossly contaminated and could have been a major source of contamination of the milk. However, the tanks were better than the milking machines in this respect.

Outlet plugs (Fig. 7) sometimes become heavily contaminated with bacteria, although they appear to be clean, because of the unsatisfactory way in which the rubber bung is attached to the metal shaft. Table XI shows that 9% of plug rinses had total bacterial counts of $>1 \times 10^9$ cfu per plug. Once a month the bung should be immersed in boiling water for 2 min; conduction of heat to the junction of the shaft and the bung where bacteria

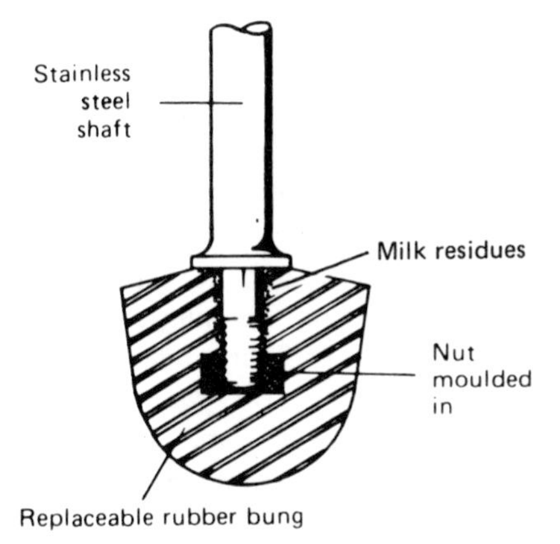


FIG. 7. Section of a bulk tank outlet plug, showing site of milk residues. (From: Cousins and McKinnon, 1979.)

may accumulate effectively disinfects the bung. These outlet plugs are peculiar to UK tanks, but the alternative to plugs, outlet cocks, are also difficult to clean and milk residues tend to accumulate in the tap seatings.

*Bucket Milking Equipment*

The bacterial content of these machines, which have to be cleaned by hand usually by brushing in a warm detergent–disinfectant solution, can vary just as widely as that of pipeline machines. Milk residues and milkstone tend to build up, because of the difficulty of cleaning the complicated teat cup clusters and the time needed to brush them effectively. Consequently, chemical disinfection is ineffective, and this is reflected in the high rinse counts reported by many workers. Flushing teat cup clusters with hot solutions, wet storage (or lye storage), immersion cleaning and treatment with boiling water have been advocated to improve the bacteriological condition of bucket milking equipment.

*Milk Cans*

If cans are not effectively cleaned and are still moist when the lids are put on them, bacterial multiplication in the moisture on the interior surfaces can result in very high bacterial counts. This source of contamination of the milk put into the cans is all the more serious because the milk is usually only water-cooled, and the bacteria from the cans can multiply rapidly in it. Cans may also be a source of *Bacillus cereus* spores and other types of thermoduric bacteria in milk.

**Types of Micro-Organisms on Milking Equipment Surfaces**

As might be expected the groups and genera of micro-organisms (able to form colonies on plate count agars) that have been found on milk contact surfaces of equipment are similar to those found in fresh raw milk (Table II). However, to be detected in bacteriological rinses by picking colonies from plates of the rinses, specific types of micro-organism must be present in appreciable proportions (> 5 %) on the equipment, and this suggests that they are protected in some way from cleaning and disinfection procedures and are subsequently able to proliferate between milkings. Selective methods can, of course, be employed to detect some minority groups of organisms, e.g. coliforms and bacterial spores.

Certain species, notably mastitis pathogens, have not been reported as forming any appreciable part of the microflora of milking equipment, although large numbers of streptococcal and staphylococcal mastitis organisms can be present in milk passing through the equipment. They can certainly survive long enough to be transferred, on a liner in a teat cup cluster used to milk a cow having an infected quarter, to the non-infected quarters of other cows. These organisms may also be transferred on udder washing cloths and on milkers' hands, but they are probably unable to multiply on surfaces of milking equipment between milkings.

The temperatures of solutions used for cleaning and disinfection influence the microflora. Pipeline milking machines subjected to ABW cleaning properly applied are, in effect, pasteurised and thus only thermoduric organisms should survive. Application of very hot detergent–disinfectant solutions has a similar effect. In practice, pasteurising temperatures are not always achieved, but heat resistant types, asporogenous Gram-positive rods (probably *Microbacterium* spp.), micrococci, streptococci and *Bacillus* spp. predominate following hot cleaning treatments, and Gram-negative rods including coliforms, which are heat labile, are relatively infrequent. If the microflora recovered by rinsing is predominantly heat labile, then it is evident that parts of the machine have not been adequately heated, and this provides guidance on the corrective measures required.

Use of solutions at lower temperatures (40–50 °C) might be expected to permit development of a heterogeneous microflora. Yet it appears that in many pipeline machines, the microflora is relatively restricted with one particular group or, at most, two of the following groups predominating: micrococci, streptococci, Gram-negative rods and asporogenous Gram-positive rods. In any one machine, the microflora can be consistent or can vary from time to time. At present, no explanation can be offered for these observations. Druce and Thomas (1972) and Thomas and Thomas (1977*a*, *b*, *c*, *d*; 1978*a*, *b*) have reviewed comprehensively the results of many workers on the bacterial content of pipeline milking machines, farm bulk tanks and bucket milking machines, with particular reference to thermoduric, psychrotrophic and coliform organisms.

Teat cup clusters of bucket milking machines showing a build-up of milk residues and milkstone were often, but not always, found to have a high thermoduric bacterial count. These organisms tend to be less prevalent in pipeline milking machines, perhaps because in-place cleaning is rather more effective in keeping surfaces superficially clean. A study of the thermoduric and psychrotrophic bacterial content of milking equipment (Mackenzie, 1973) showed that the incidence of thermoduric organisms in pipeline milking machines was slightly lower than that of psychrotrophs, none of the machines having thermoduric counts exceeding $1 \times 10^7$ cfu m$^{-2}$ as compared with 7·5% having psychrotrophic counts above this level.

The total bacterial content of farm bulk tanks is lower than that of milking machines, and the thermoduric bacterial content is very low, $<1 \times 10^5$ cfu m$^{-2}$; this is because most thermoduric bacteria cannot multiply in the cold environment of the tanks. The proportion of Gram-negative rods and psychrotrophs in tanks is, however, much higher than in milking machines.

In the UK there has been a tendency towards an increase in the level of total bacterial contamination in tanks and plugs and Mackenzie's (1973) results indicate that the content of psychrotrophs in tanks has also increased.

In view of the increasing interest and concern in the numbers and types of psychrotrophs in raw milk, there is a surprising lack of information on the incidence of these organisms on milking equipment, and the influence of cleaning and disinfection treatments and other factors on their survival on milk contact surfaces. However, it is fair to assume that where the total bacterial content of equipment is low, that of specific, undesirable types is also likely to be low. The British Standards Institution (1975) (BS 5226) details cleaning and disinfecting methods for achieving satisfactory bacteriological cleanliness in pipeline milking machines and in bulk tanks.

## THE INFLUENCE OF STORAGE AND TRANSPORT ON THE MICROFLORA OF RAW MILK

After production, milk is stored either in cans or in a bulk tank to await collection and subsequent delivery to a collection centre, a processing dairy or a manufacturing creamery.

**Can Collection**

In temperate climates, milk is usually collected once a day and, normally, is water-cooled to as low a temperature as possible. This depends on the method of cooling and the temperature of the available water supply; in summer part of the daily milk yield may have to be stored for 14–18 h at 20–25 °C. Where ambient temperatures frequently exceed 25 °C and, sometimes 30 °C, un-refrigerated milk is often collected twice daily because of the rapidity of bacterial multiplication and the consequent high risk of souring or spoilage of milk held at such temperatures for more than about 6 h. On arrival at its destination, can milk is either cooled to ≤5 °C and stored for no more than 24 h before processing or manufacture, heat treated and then cooled before storage, or it is used immediately.

**Bulk Collection**

With this system, milk is most commonly refrigerated immediately after production, either by means of an in-line cooler followed by storage in an insulated tank, or in a tank equipped with a refrigeration system. Collection of the milk which may be daily, on alternate days, every third day or, more rarely, at longer intervals, is by means of insulated transport tankers each picking up milk from several farms. Thus, there is a risk that an undetected faulty consignment from one farm may spoil a whole load of milk. Tanker drivers are normally authorised to refuse to collect milk which is tainted, appears to be abnormal or is above a specified temperature, e.g. 7 °C. Bulk collection schemes may include requirements and specifications concerning the design and performance of farm bulk tanks, the rate at which the milk is to be cooled, the maximum temperature at which it is to be stored, the frequency of collection, and the cleanliness of the transport tanker (Hoyle, 1979).

In some countries, water-cooled milk is stored in insulated farm tanks and is collected daily or, at the height of the production season, twice daily. This type of bulk collection system is used only for manufacturing grade milk.

On arrival at its destination, farm refrigerated milk, if it is not used immediately, is transferred to insulated storage tanks or silos (see Fig. 8) of up to 30 000 gal. capacity, where it is stored until heat treated for processing or manufacture; in some countries, this may be up to 4 days after collection for daily, or alternate day, collected milk. In The Netherlands, where some refrigerated milk is collected every third day, such milk, unless used immediately, is 'thermised', i.e. heated to *c*. 63 °C for 15 s on arrival at the dairy; this is sufficient to kill most of the psychrotrophs, so that after cooling to 5–6 °C, the milk can then be stored for 2–3 days before use.

FIG. 8. These insulated storage silos are typical of those employed to store ex-farm bulk milk prior to processing into cheese or other milk products. The larger silos hold around 30 000 gal. of milk at 3–5 °C, the smaller ones around 25 000 gal. Reproduced by courtesy of the Milk Marketing Board, Davidstow.

## Bacterial Multiplication in Stored Milk

The temperature and duration of storage, the numbers and types of bacteria in the milk and, to a lesser extent, the natural inhibitory systems in the milk, all influence the increase in bacterial numbers which occurs in stored milk. Because of the wide variation in the initial microflora, and in the conditions under which milk is stored, only generalisations can be made concerning changes in the microflora of milk occurring during storage and transport.

The temperature of storage is probably the most important factor, and Fig. 9 illustrates its likely effect on milk of rather poor quality (having an initial total count of 50 000 cfu $ml^{-1}$) and, from the curves, the importance of cooling, if milk is to be kept for more than about 12 h is evident; it is assumed that adverse effects start to become apparent in milk when the count approaches $1 \times 10^7$ cfu $ml^{-1}$. Milk delivered in cans is often checked on arrival for these effects and, if tainted or having an acidity above specified limits, the milk is down-graded or rejected.

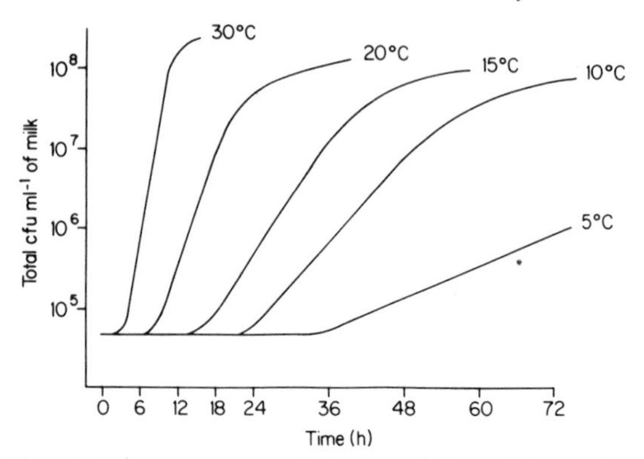


FIG. 9. The effect of milk temperature on the increase in bacterial count in raw milk having an initial total count of 50 000 cfu ml$^{-1}$.

The spoilage organisms that become predominant at 25–30 °C are mainly streptococci and coliforms, and both types increase the acidity of the milk. Gram-negative rods, other than coliforms, and micrococci (including staphylococci) will also multiply unless, or until, any developed acidity becomes inhibitory to them. There are wide variations between individual milk supplies because of variations in the initial microflora, and unpleasant taints may mask the 'clean' acid odour associated with a predominance of streptococci. Between about 15 °C and 25 °C, Gram-negative rods sometimes outnumber streptococci but, in general, the effects are similar to those at higher temperatures, except that they are delayed for some hours. Many species of Gram-negative rods multiplying in milk do not produce any noticeable effects even when their numbers reach or exceed $1 \times 10^7$ cfu ml$^{-1}$; thus, at lower storage temperatures, e.g. 10 °C or less, the milk may appear normal for 2–3 days, although it is likely that considerable bacterial multiplication will have occurred, unless external contamination of milk during production is minimal.

Druce and Thomas (1968) have reviewed the effects of pre-incubation for specified periods and at specified temperatures on the results of bacteriological tests applied to milk samples before and after pre-incubation. In general, saprophytic bacteria multiply most readily under the commonly used pre-incubation conditions, i.e. 12–22 °C for 16–24 h. Micro-organisms comprising the udder microflora, having an optimum growth temperature of 37 °C, multiply slowly if at all during pre-incubation, so that a 100-fold increase (for example) in colony count during storage is indicative of bacterial contamination from sources outside the udder. However, pre-incubation is not infallible in this respect, because the saprophytic thermoduric coryneforms and micrococci derived from milking equipment surfaces do not multiply in raw milk within 24 h at pre-incubation temperatures. On the other hand, some types of coliforms from within the udder are able to do so.

There is no doubt that most milk produced with the minimum of external bacterial contamination shows much smaller increases in bacterial numbers on storage over a wide temperature range as compared with heavily contaminated milk. The natural inhibitory properties of milk play some part in this effect, and bacterial multiplication is delayed even in moderately contaminated milk for 2–3 h at 30 °C, and for longer periods at lower temperatures (Fig. 9).

**Refrigerated Storage of Raw Milk**

Refrigeration, by delaying bacterial multiplication, masks the effects of unhygienic production conditions which, in poorly cooled milk in warm weather, result in souring and other obvious forms of spoilage of milk.

In many dairying regions, bulk collection of refrigerated milk accounts for most, or all, of the milk produced and the numbers and types of the initial psychrotrophic microflora and their activities during storage are becoming increasingly important.

Alternate day (AD) collection is a common practice, so that four successive additions of milk are made to the tank, and about a quarter of the milk is, therefore, 2 days old at collection. Studies of AD collection between 1950 and 1970 showed that it was not detrimental to the bacteriological quality of the milk up to the point of leaving the farm, provided that the milk added to the tank after each milking was cooled rapidly to $\leq 4$ °C (Thomas *et al.*, 1971). There is probably a slight increase in the numbers of psychrotrophs, but this is insignificant because of dilution resulting from the repeated addition of fresh milk. However, storage at 5–7 °C and a high psychrotroph count in the fresh milk, is likely to lead to a marked increase in count at the time of collection. Results of investigations on the bacteriological quality of AD collected milk have been reviewed by Thomas and Druce (1971).

The initial total viable count of raw milk is of little value for predicting its count after refrigerated storage. Samples from farm bulk tanks containing milk from two milkings, and taken shortly after the addition of the second milking, were stored at 5 °C and standard plate counts determined at daily intervals. The wide range of responses to storage is shown in Table XII; some samples showed only small changes after 4 days and others 3- to 8-fold increases in 2 days. As might be expected, in the four samples in which

TABLE XII

TOTAL BACTERIAL COUNTS OF INDIVIDUAL FARM BULK TANK MILKS STORED AT 5 °C

| Farm | $cfu\ ml^{-1}$ of milk after storage for: 0 days | 2 days | 3 days | 4 days |
|---|---|---|---|---|
| A | 5 800 | 3 300 | 7 900 | 14 000 |
| B | 14 000 | 10 000 | 11 000 | 70 000 |
| C | 14 000 | 10 000 | 710 000 | 15 000 000 |
| D | 28 000 | 83 000 | 2 800 000 | 18 000 000 |
| E | 62 000 | 400 000 | 9 500 000 | 41 000 000 |
| F | 170 000 | 110 000 | 110 000 | 130 000 |
| G | 240 000 | 1 800 000 | 8 900 000 | 17 000 000 |

bacterial multiplication had been most rapid, the psychrotrophic count and the total count were similar after 3 days.

**The Effects of Refrigerated Transport and Subsequent Storage**

Farm collection tankers are insulated, and the temperature of the large volumes of milk they carry is unlikely to rise much during transport and is, therefore, dependent on the temperature of the farm tank milk when collected. Much milk arrives at its destination at $\leq 5$ °C, but surveys of milk arriving at manufacturing plants have shown that tanker milk temperatures have frequently been in the range 6–9 °C. Often there are no facilities for cooling incoming milk to $\leq 4$ °C, should this be necessary, before it is put into bulk storage. Bacterial counts on samples taken from farm collection tankers and bulk storage tanks and subsequently held in the laboratory at temperatures ranging from 5–10 °C, show the importance of keeping the temperature as low as possible if bacterial multiplication in the stored milk is to be minimised. The results of storing samples of daily collected milk having a mean initial psychrotroph count of $1 \times 10^4$ cfu ml$^{-1}$ are shown in Fig. 10. Even at 5°C, the count exceeded $1 \times 10^6$ cfu ml$^{-1}$ in 3 days (Cousins *et al.*, 1977); other workers have reported greater increases in milk samples taken from bulk storage tanks prior to laboratory storage. Standard plate counts of 'commingled' milk from manufacturing plant silo tanks were *c.* $1 \times 10^7$ cfu ml$^{-1}$ after one day's storage (LaGrange, 1979), but milk in some of the incoming tankers had SPC of $>1 \times 10^6$ cfu ml$^{-1}$. Muir *et al.* (1978) recorded counts ranging from $10^4$–$5 \times 10^6$ cfu ml$^{-1}$ for silo tank milks, some of which related to AD collected milk stored for 24 h at the creamery. The total psychrotroph and coliform counts of tanker milk

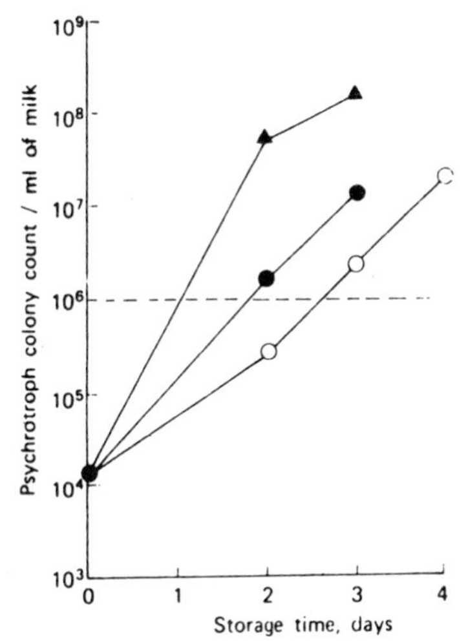


FIG. 10. Psychrotroph counts (milk, agar, 5°C/10 days) of farm tanker milk showing increases during storage at 5°C (○), 7°C (●) and 10°C (▲); means of eight samples.

arriving at creameries have been found to be high compared with those of milk sampled from individual farm tanks, but there was little difference in the thermoduric counts (Thomas, 1974*a*).

It is not clear to what extent bacterial contamination from transport tankers, hoses, pumps and, where fitted, meters and automatic samplers contributes to the increased bacterial content of milk arriving in tankers for transfer to bulk storage at processing dairies. Because of the large volumes of milk involved, e.g. 10 000 litres, any significant increase in count per ml of milk indicates very heavy contamination of the milk contact surfaces. Some increase in count may occur as a result of pumping, which could disrupt clumps and chains of bacteria in the milk, but there is no clear evidence concerning the extent of any such increase. One or two unusually heavily contaminated individual farm supplies will, of course, influence a whole tanker load. For milk that is more than about 12 h in transit, and that has a high content of psychrotrophs, multiplication of these organisms may also play a part if the milk temperature is $>4$°C.

During storage, multiplication of psychrotrophs derived from bulk handling equipment will contribute to the bacterial content of stored milk. If fresh milk is added to storage tanks already containing milk stored for 24 h or longer, and the mixed milk is held over for a further period, then again bacterial counts are likely to reach undesirable levels more quickly than if the milk was put into a clean empty tank.

To ensure that raw milk in storage does not seriously deteriorate in bacteriological quality, the temperature of the incoming milk should be monitored and kept below 5 °C, farm tankers and their ancillary equipment, storage and silo tanks should be effectively cleaned and disinfected as soon as they are emptied, and prolonged storage of raw milk and the practice of topping up partially emptied silo tanks should be avoided.

Careful management and control of stocks of milk would be assisted by daily monitoring of their bacteriological quality; any faults or problems revealed could then be rectified. For this purpose, a new method for the direct enumeration of bacteria in milk, taking only 20 min and correlating well with the plate count (Pettipher *et al.*, 1980), could be useful.

**Types of Bacteria in Stored Milk**

Generally within 2–3 days after transfer from transport tankers, the microflora of the milk is dominated by psychrotrophs, whereas the thermoduric microflora does not increase, and changes little in composition. Spores of psychrotrophic *Bacillus* spp. will be present, but germination of spores and outgrowth is unlikely to occur to any significant extent at the temperatures, and for the time, raw milk is stored before use.

The predominant psychrotrophic genera and species will vary, but will be derived from those present in the raw milk initially (Table IV). *Pseudomonas* spp. are the most frequently reported psychrotrophs in stored milk with *Ps. fluorescens* the most common although, as pointed out by Law (1979) in his review, this may be due to the ease with which this species may be tentatively identified, rather than to its true distribution.

Pasteurisation will destroy the psychrotrophic bacteria, but not necessarily the products of their metabolism or their enzymes. Small scale experiments have shown that heat resistant bacterial enzymes can adversely affect product yield and quality. Although, on a factory scale demonstration of these effects is not clear cut, this is probably because sufficiently detailed studies on the bacteriological quality of raw milk immediately before processing are not at present available. However, good handling practices will avoid high bacterial counts and the possible presence of undesirable bacterial enzymes or other effects.

## REFERENCES

Akam, D. N. (1979) Description and performance of components. In: *Machine Milking*, Thiel, C. C. and Dodd, F. H. (Eds.), NIRD, Reading.

Anon. (1979) *Veterinary Record*, **105**, 294.

BENHAM, C. L. and EGDELL, J. W. (1970) *J. Soc. Dairy Technol.*, **23**, 91.
BERGÈRE, J.-L. (1979) *Revue Laitière Française*, **378**, 19.
BRAMLEY, A. J. (1975) Infection of the udder with coagulase negative micrococci and *Corynebacterium bovis*. In: *Proc. Seminar on Mastitis Control*, Dodd, F. H., Griffin, T. K. and Kingwill, R. G. (Eds.), International Dairy Federation Doc. No. 85, Brussels.
BRAMLEY, A. J., KING, J. S. and HIGGS, T. M. (1979) *Brit. Vet. J.*, **135**, 262.
BRAMLEY, A. J. and NEAVE, F. K. (1975) *Brit. Vet. J.*, **131**, 160.
BRITISH STANDARDS INSTITUTION (1975) *Recommendations for cleaning and sterilization of pipeline milking machine installation*, BS 5226, BSI, London.
BRYAN, F. L. (1969) Infections due to miscellaneous micro-organisms. In: *Food Infections and Intoxications*, Riemann, H. (Ed.), Academic Press, New York and London.
CARREIRA, D. F. C., CLEGG, L. F. L., CLOUGH, P. A. and THIEL, C. C. (1955) *J. Dairy Research*, **22**, 166.
CARROLL, E. J., JAIN, N. C., SCHALM, O. W. and LASMANIS, J. (1973) *Am. J. Vet. Research*, **34**, 1143.
COUSINS, C. M. (1963) *J. Appl. Bacteriol.*, **26**, 376.
COUSINS, C. M. (1972) *J. Soc. Dairy Technol.*, **25**, 200.
COUSINS, C. M. (1978) Milking techniques and the microbial flora of milk, *XXth International Dairy Congress, Paris:* Congress lecture.
COUSINS, C. M. and MCKINNON, C. H. (1979) Cleaning and disinfection in milk production. In: *Machine Milking*, Thiel, C. C. and Dodd, F. H. (Eds.), NIRD, Reading.
COUSINS, C. M., SHARPE, M. E. and LAW, B. A. (1977) *Dairy Industries Internat.*, **42**, 12.
DAVIES, F. L. and WILKINSON, G. (1973) *Bacillus cereus* in milk and dairy products. In: *The Microbiological Safety of Food*, Hobbs, B. C. and Christian, J. H. B. (Eds.), Academic Press, London and New York.
DRUCE, R. G. and THOMAS, S. B. (1968) *Dairy Sci. Abstr.*, **30**, 291.
DRUCE, R. G. and THOMAS, S. B. (1972) *J. Appl. Bacteriol.*, **35**, 253.
FORD, J. E. (1974) *Brit. J. Nutrition*, **31**, 243.
GOUDKOV, A. V. and SHARPE, M. E. (1965) *J. Appl. Bacteriol.*, **28**, 63.
HAHN, G., HEESCHEN, W. and TOLLE, A. (1970) *Kieler Milchwirtschaftliche Forschungsberichte*, **22**, 335.
HOYLE, J. B. (1979) Milk cooling equipment. In: *Machine Milking*, Thiel, C. C. and Dodd, F. H. (Eds.), NIRD, Reading.
INTERNATIONAL DAIRY FEDERATION (1974) *Bacteriological quality of cooled bulk milk*. Doc. No. 83, IDF, Brussels.
INTERNATIONAL DAIRY FEDERATION (1980) *Factors influencing the bacteriological quality of raw milk*. Doc. No. 120, IDF, Brussels.
JACKSON, H. and CLEGG, L. F. L. (1966) *Canadian J. Microbiol.*, **12**, 429.
JOHNS, C. K. (1962) The coliform count of raw milk as an index of udder cleanliness. *XVIth International Dairy Congress*, Copenhagen, **C**, 365.
JOHNS, C. K. (1971) *J. Milk and Fd Technol.*, **34**, 173.
JUFFS, H. S. (1973) *J. Appl. Bacteriol.*, **36**, 585.
KINGWILL, R. G., DODD, F. H. and NEAVE, F. K. (1979) Machine milking and mastitis. In: *Machine Milking*, Thiel, C. C. and Dodd, F. H. (Eds.), NIRD, Reading.

LaGrange, W. S. (1979) *J. Fd Protection*, **42**, 599.
Lander, K. P. and Gill, K. P. W. (1979) *Veterinary Record*, **105**, 333.
Law, B. A. (1979) *J. Dairy Research*, **46**, 573.
Lück, H. (1972) *Dairy Sci. Abstr.*, **34**, 101.
Mackenzie, E. (1973) *J. Appl. Bacteriol.*, **36**, 457.
McKinnon, C. H. and Cousins, C. M. (1969) *J. Soc. Dairy Technol.*, **22**, 227.
Mikolajcik, E. M. (1979) *Cultured Dairy Products J.*, **14**, 6.
Muir, D. D., Kelly, M. E., Phillips, J. D. and Wilson, A. G. (1978) *J. Soc. Dairy Technol.*, **31**, 137.
Murphy, J. M. (1943) *Cornell Veterinarian*, **33**, 48.
Olsen, S. J. (1975) A mastitis control system based upon extensive use of mastitis laboratories. In *Proc. Seminar on Mastitis Control*, Dodd, F. H., Griffin, T. K. and Kingwill, R. G. (Eds.), International Dairy Federation Doc. No. 85, Brussels.
Olsen, J. C., Casman, E. P., Baer, E. F. and Stone, J. E. (1970) *Appl. Microbiol.*, **20**, 605.
Palmer, J. (1977) *Irish J. Fd Sci. and Technol.*, **I**, 57.
Panes, J. J., Parry, D. R. and Leech, F. B. (1979) *Report of a survey of the quality of farm milk in England and Wales in relation to EEC proposals*, Ministry of Agriculture, Fisheries and Food, London.
Pettipher, G. L., Mansell, R., McKinnon, C. H. and Cousins, C. M. (1980) *Appl. and Environmental Microbiol.*, **38**, 423.
Read, R. B. and Bradshaw, J. G. (1966) *J. Dairy Sci.*, **49**, 202.
Reiter, B. (1978) *J. Dairy Research*, **45**, 131.
Reiter, B. and Bramley, A. J. (1975) Defence mechanism of the udder and their relevance to mastitis control. In: *Proc. Seminar on Mastitis Control*, Dodd, F. H., Griffin, T. K. and Kingwill, R. G. (Eds.), International Dairy Federation Doc. No. 85, Brussels.
Ridgeway, J. D. (1955) *J. Appl. Bacteriol.*, **18**, 374.
Robinson, D. A., Edgar, W. J., Gibson, G. L., Matchett, A. A. and Robertson, L. (1979) *Brit. Medical J.*, **1**, 1171.
Russell, M. W., Brooker, B. E. and Reiter, B. (1977) *Research in Vet. Sci.*, **20**, 30.
Schalm, O. W., Lasmanis, J. and Carroll, E. J. (1964) *Am. J. Vet. Research*, **25**, 83.
Schalm, O. W., Carroll, E. J. and Jain, N. C. (1971) *Bovine Mastitis*. Lea & Febiger, Philadelphia.
Schalm, O. W., Lasmanis, J. and Jain, N. C. (1976) *Am. J. Vet. Research*, **37**, 885.
Scott, W. M. and Minett, F. C. (1947) *J. Hygiene*, **45**, 159.
Sinell, H. J. (1973) Food infections from animals. In: *The Microbiological Safety of Food*, Hobbs, B. C. and Christian, J. H. B. (Eds.), Academic Press, London and New York.
Smith, J. (1934) *J. Comparative Pathol.*, **47**, 125.
Tarry, D. W. (1979) *State Vet. J.*, **34**, 250.
Taylor, P. R., Weinstein, W. M. and Bryner, J. H. (1979) *Am. J. Medicine*, **66**, 779.
Thiel, C. C., Cousins, C. L., Westgarth, D. R. and Neave, F. K. (1973) *J. Dairy Research*, **40**, 117.
Thomas, S. B. (1974*a*) *J. Soc. Dairy Technol.*, **27**, 180.

THOMAS, S. B. (1974*b*) *Dairy Industries*, **39**, 237, 279.
THOMAS, S. B. and DRUCE, R. G. (1971) *Dairy Sci. Abstr.*, **33**, 339.
THOMAS, S. B. and DRUCE, R. G. (1972) *Dairy Industries*, **37**, 593.
THOMAS, S. B. and THOMAS, B. F. (1975*a*) *Dairy Industries*, **40**, 338.
THOMAS, S. B. and THOMAS, B. F. (1975*b*) *Dairy Industries*, **40**, 478.
THOMAS, S. B. and THOMAS, B. F. (1977*a*) *Dairy Industries Internat.*, **42**(4), 7.
THOMAS, S. B. and THOMAS, B. F. (1977*b*) *Dairy Industries Internat.*, **42**(5), 18.
THOMAS, S. B. and THOMAS, B. F. (1977*c*) *Dairy Industries Internat.*, **42**(7), 19.
THOMAS, S. B. and THOMAS, B. F. (1977*d*) *Dairy Industries Internat.*, **42**(11), 25.
THOMAS, S. B. and THOMAS, B. F. (1978*a*) *Dairy Industries Internat.*, **43**(5), 17.
THOMAS, S. B. and THOMAS, B. F. (1978*b*) *Dairy Industries Internat.*, **43**(10), 5.
THOMAS, S. B., HOBSON, P. M., BIRD, E. R., KING, K. P., DRUCE, R. G. and COX, D. R. (1962) *J. Appl. Bacteriol.*, **25**, 107.
THOMAS, S. B., DRUCE, R. G. and KING, K. P. (1966) *J. Appl. Bacteriol.*, **29**, 409.
THOMAS, S. B., DRUCE, R. G., PETERS, G. J. and GRIFFITHS, D. G. (1967) *J. Appl. Bacteriol.*, **30**, 265.
THOMAS, S. B., DRUCE, R. G. and JONES, M. (1971) *J. Appl. Bacteriol.*, **34**, 659.
TITTERTON, M. and OLIVER, J. (1979) *Rhodesian J. Agric. Research*, **17**, 19.
UNDERWOOD, H. M., MCKINNON, C. H., DAVIES, F. L. and COUSINS, C. M. (1974) *XIXth International Dairy Congress*, **1E**, 373.
VAKIL, J. R., CHANDAN, R. C., PARRY, R. M. and SHAHANI, K. M. (1969) *J. Dairy Sci.*, **52**, 1192.
WEIR, J. and BARBOUR, D. (1950) *Vet. Record*, **62**, 239.
WERNER, S. B., HUMPHREY, C. L. and KAMEI, I. (1979) *Brit. Medical J.*, **2**, 238.
WESTGARTH, D. R. (1975) Interpretation of herd bulk milk cell counts. In: *Proc. Seminar on Mastitis Control*, Dodd, F. H., Griffin, T. K. and Kingwill, R. G. (Eds.), International Dairy Federation Doc. No. 85, Brussels.
WHEELOCK, J. V., ROOK, J. A. F., NEAVE, F. K. and DODD, F. H. (1966) *J. Dairy Research*, **33**, 199.
WILSON, C. D. and KINGWILL, R. G. (1975) A practical mastitis control routine. In: *Proc. Seminar on Mastitis Control*, Dodd, F. H., Griffin, T. K. and Kingwill, R. G. (Eds.), International Dairy Federation Doc. No. 85, Brussels.
WITTER, L. D. (1961) *J. Dairy Sci.*, **44**, 983.
ZARKOWER, A. and SCHEUCHENZUBER, W. J. (1977) *Cornell Veterinarian*, **67**, 404.

# 5

# The Microbiology of Market Milk

F. E. NELSON
*Emeritus Professor of Food Science,*
*Department of Dairy and Food Science, University of Arizona, Tucson, USA*

In this chapter, the concern will be with milk for consumption as fluid milk, primarily as the pasteurised product. The quality of the incoming raw milk is of great importance, since subsequent handling and processing to produce a final product of satisfactory acceptability cannot overcome deficiencies in the raw milk supply. Although primary concern is for cow's milk, milk from other mammals may be very important in some areas.

Fluid milk is a highly perishable commodity. Milk also has special importance as a component of the diets of the young and the old, as both of these groups tend to have greater vulnerability to adverse dietary factors than do people in the intermediate age groups. For both of these reasons, greater emphasis has been placed upon the quality of milk than upon the quality of most other food groups. In some instances this concern has been carried to the point where basic requirements have been superseded by primarily aesthetic considerations. Some thought must be given to what may be considered good manufacturing and handling practices within a given environment. Practices considered highly desirable, or even mandatory, under one combination of social, technological, nutritional and economic conditions could be completely impractical, at least in the early phases of programme development, in another situation. Certainly, where the need is great for the nutritional advantages of milk and milk products, adoption of either marginal or unnecessary requirements would be of very questionable value.

Micro-organisms are of concern in market milk because of the following points.

1. They are very important causes of spoilage, and a spoiled product is

a nutritionally unavailable product and an economic loss to producer, processor and consumer.

2. Milk may be a carrier of micro-organisms, or the metabolic products of micro-organisms, that can cause human illness. Proper production conditions to minimise contamination and adequate treatment to inactivate such potentially harmful organisms are essential.
3. Quantitative and qualitative determinations of microbial populations or products thereof can be used, along with other criteria, such as some chemical determinations, to assess the adequacy of treatment and protection employed.
4. Certain micro-organisms or their metabolic products, such as enzymes, may be used to bring about desirable modifications of a product.

## REGULATORY CONTROL OF MILK SUPPLIES

Most of the fluid milk consumed in the world, and essentially all of that consumed in areas of high technological development, must satisfy some governmental regulations. These regulations include bacteriological standards for the product and specify levels of operation which are considered important in satisfying bacteriological and other requirements. These ordinances and regulations are implemented by sanitarians going to the producing farms and processing plants to evaluate conditions under which the milk is produced and handled, and to take appropriate samples for laboratory examination. Governmental ordinances and regulations are considered to be the minimum standards which the industry should meet. Industry frequently imposes higher standards, both because higher standards provide a margin for operation, increasing the probability that regulatory standards will be met consistently, and because higher standards may help to ensure good keeping quality, better flavour and other commercially desirable attributes of the product. Trade groups, such as the Milk Industry Foundation in the United States, frequently work closely with government regulatory groups in developing ordinances and regulations which will be governmentally effective, and that the industry can implement most effectively. Organisations such as the International Dairy Federation, the Food and Agriculture Organisation of the United Nations and the National Conference on Interstate Milk Shipments in the United States also provide an input of this type. Laboratory methods and interpretations are considered by such groups as the American Public

Health Association, the Association of Official Analytical Chemists, the International Association of Milk, Food and Environmental Sanitarians, the American Dairy Science Association and similar groups made up, to a significant degree, of scientifically trained individuals. These groups have combined to provide the background for more regulation of the milk industry than for most other food industries, but also to help provide a product which has an extremely high level of safety and nutritional quality for the consumer.

**Principles of Effective Control**

An effective milk control programme depends upon: (1) establishment of standards; (2) use of effective enforcement; and (3) education of producers, processors, distributors and consumers.

Effective standards must apply to all aspects of production, processing and distribution. They relate to animal and human health, to equipment design and fabrication, to construction of buildings involved, to proper handling of the product at all stages (including adequate heat treatment) and to the bacteriological, chemical and other standards which the product must meet. The standards must be adequate to provide an acceptable product, without imposing unrealistic requirements on those involved in any aspect of placing the product before the final consumer.

Standards are ineffective unless enforced. This is an area in which the sanitarian, qualified by training and experience to interpret the regulations under which he works, is vital to the success of the programme. The laboratory which provides the dependable quantitative and qualitative data to ensure that the product satisfies, at least, minimum standards is an essential link in enforcement. Supervisory personnel, both those intimately involved and those in the echelons which provide funds, judicial support and liaison with other agencies, must be supportive of the programme. Without such support, adequate enforcement becomes virtually impossible.

Provisions must be made for penalties for non-compliance with the standards established by ordinance or regulation. Suspension or revocation of a permit or license may be used, resulting in the inability of a manufacturer to market his product. Monetary fines may be used under some circumstances. Degrading, which either prevents sale of the product or which requires changing the label to a lower grade less acceptable to the buyer, is used in many cases. Consistency of use of the procedure chosen and adequate support from the judiciary are essential for a viable programme.

Education is a necessary part of any regulatory programme. Producers, processors and distributors will comply with requirements much more readily and intelligently if they know why the requirement is being imposed. The consumer will be more apt to support the programme and be less apt to make unnecessary demands, if he or she has adequate information available. All too frequently regulatory officials, producers, processors, distributors and consumers become emotionally involved because of inadequate knowledge or biased or misleading interpretations.

**Regulatory Agencies**

The fluid milk industry has changed over the years from one in which a few quarts of raw product were marketed literally 'over the fence' in a very local environment, into one in which raw milk frequently is shipped a thousand miles or more in large tankers to be pasteurised and packaged in central plants handling thousands of gallons of milk a day. The packaged product may in turn be shipped hundreds of miles to the final outlet and consumer. Large-scale programmes to control brucellosis and tuberculosis among milk-producing animals, and much greater awareness of the economic, as well as the disease-related, aspects of mastitis have reduced greatly the presence in raw milk of micro-organisms which might cause human illness. The shift from raw milk to pasteurised or sterilised milk has made great differences in the incidence of milk-borne disease, and in the keeping quality of the product. The extensive use of good refrigeration on the farm, and all along the processing, distribution and consumption line, has helped alter the pattern of operation. Today, dried products, such as milk-solids-not-fat, are prepared in one country and shipped many miles to be reconstituted into fluid products which are processed, packaged and consumed far from the point of origin of the components. These changes have had a marked impact on regulatory patterns.

From a situation where regulation was almost exclusively local, national and international, considerations rank large in the regulation of milk supplies. However, although the substance of the ordinances, regulations and standards now tends to originate at the national and international level, actual enforcement is usually at a relatively local level. With proper co-ordination between jurisdictions, results from one area are increasingly accepted by other geographical areas. In the United States, this confidence has resulted from the activities of the Public Health Service (this activity is now a component of the Food and Drug Administration) and the National Conference on Interstate Milk Shipments, both of which have done much to put regulatory activities on a uniform basis. The days of multiple 'inspection' of producer and processor are not ended, but great progress in this direction has been made. Increasingly, the results of industrial 'self policing' are being accepted as a valid portion of the regulatory process, to the advantage of all concerned.

Several federal agencies are involved in regulatory activities that apply to milk. The Food and Drug Administration, through a group which was formerly in the Public Health Service, prepares a number of ordinances and associated interpretive codes which are recommended for adoption by various agencies involved in control of milk supplies. Among these ordinances are those for Grade A pasteurised milk (USFDA, 1978*a*), Grade A condensed and dry milk products and condensed and dried whey (USFDA, 1978*b*), and fabrication of single-service containers and closures for milk and milk products (USFDA, 1978*c*). Assistance is also given to states and municipalities in making ratings of milk supplies, and in certifying laboratories of state agencies for the bacteriological and chemical examination of milk. These are all activities which are helpful in promulgating uniform and effective milk control programmes. The Food and Drug Administration also formulates definitions for many dairy products, and sets standards for the control of unwholesome or adulterated milk products. In recent years, they have been particularly active in monitoring for pesticide residues, mycotoxins and antibiotic residues.

The United States Department of Agriculture has been active in programmes for the control and elimination of tuberculosis and brucellosis in dairy animals. They have also been active in setting standards for manufactured dairy products, and in conducting research relating to milk production, processing and distribution.

The Department of Defence has promulgated standards for the processing of dairy products sold to the armed forces, and has done some work on the development of products.

At the state level in the United States, either the Health Department or the Department of Agriculture (in a few instances a combination of the two) is usually responsible for promulgating and enforcing regulations for the control of milk supplies. Frequently they adopt the recommended Grade A pasteurised milk ordinance (USFDA, 1978*a*), or some modification of this ordinance which is felt to be more suited to the local situation; the state agency works closely with those county and city groups which actually carry out the details of enforcement. Where no county or city ordinance applies, the complete responsibility for control is usually assumed by the state agency. The state agency may also be involved in educational activities, such as seminars which include the local personnel.

Local regulatory activity in the United States originated primarily in eastern cities that were concerned about the quality of their milk supplies around the turn of the century. Frequently personnel were sent to areas considerable distances from the city to work with the producers and forwarders of the product. As regulations and standards of enforcement have become more uniform, local units have come to depend more and more on reciprocal arrangements between units at various levels for enforcement. In addition, as metropolitan areas have expanded across city boundaries, a tendency has developed for the county or even a group of counties to take over the local regulatory function. The extent to which one processing facility can now market its output over a wide area, commonly involving a number of governmental jurisdictions, has also contributed to the enlargement of local areas of regulatory activity.

While several grades of market milk have been recognised in the past, the present tendency is to recognise only that which can be designated as Grade A or equivalent. This has been an evolutionary process, in which an increasing output of product, which will satisfy the high standards of Grade A, has been achieved. In many areas, a significant amount of Grade A milk is used for the manufacture of milk products, particularly during seasons of high production. The return on such milk to the producer is less than the return on milk used as fluid milk, primarily because of the provisions of the various Milk Marketing Orders which apply.

In the Grade A pasteurised milk ordinance, raw milk for pasteurisation should not exceed a bacterial limit of 300 000 $ml^{-1}$ as 'commingled' milk prior to pasteurisation, and must be maintained at all times, except within 2 h after milking, at 10 °C (50 °F) or less until pasteurised. In addition, the somatic cell count shall not exceed 1 200 000 $ml^{-1}$ and the milk must come from herds meeting stringent criteria for freedom from brucellosis and tuberculosis. The pasteurised milk and milk products must be cooled to 7 °C (45 °F) or below and maintained thereat. The product must not exceed a bacterial limit of 20 000 $ml^{-1}$, a limit of 10 coliform bacteria per ml, and be negative for phosphatase and antibiotics. In actual practice, these requirements are usually met by a substantial margin. Some feel that certain of these requirements should be made more stringent. However, if they are looked upon as minimum administrative requirements that, frequently, are made more stringent by processors and by some regulatory jurisdictions, they are probably reasonable guidelines, capable of being met by reasonable carefulness.

The recommended ordinance specifies, in considerable detail, such items as equipment construction, building facilities, water supplies, operation of pasteurisation equipment and other elements that contribute to a final product that is of good nutritional properties, free from micro-organisms and other agents which may cause human illness, and of satisfactory aesthetic quality. The ordinance should be consulted for details, since the volume of material covered is too great to be presented with any degree of completeness here.

**Industry Quality Control Programmes**

The milk industry undertakes quality control programmes both to ensure that the requirements of governmental regulations are satisfied adequately, and to provide greater assurance that the products which they market will satisfy consumers. Special attention is given to good keeping quality, and to freedom from defects of either flavour or physical nature. Appropriate laboratory control and trained personnel, such as in-plant sanitarians and fieldmen who work with producers, are essential components of a good programme. To an increasing degree, trained personnel work with those involved in distribution and with those who utilise the products in food service, such as restaurants, hospitals and school lunch programmes, to maintain the quality of the product until it is consumed.

Some larger processors have their own laboratories and personnel to carry out their quality control programmes. Some rely, to a degree, on outside consultants to monitor their programmes. In some instances, a number of processors have gone together to establish a unit or units which will provide the desired services.

Producer co-operatives frequently provide the field service personnel who work with the producers. These people will make farm visits to assist with a variety of problems related to the production of quality milk, as well as frequently providing information on such items as feeding and management. The processor quite commonly has little direct contact with the producer, relying on the activity of the producer co-operative to obtain the quality of milk needed. The local government regulatory personnel work closely with the fieldmen of the producer co-operative in providing solutions to common problems.

## THE PRODUCER MILK SUPPLY

While delivery of a suitable raw material is an obligation of the producer, the processor must determine that this obligation is being fulfilled. A variety of tests will usually be made on incoming milk. Off-odours and abnormal temperatures will be checked at the receiving platform, where samples will also be taken for examination in the laboratory.

Estimations of 'total' numbers of bacteria are made most acceptably by the plate count with incubation at 28–32 °C. Standard methods for the examination of dairy products (APHA, 1978) specify 32 °C for 48 ± 3 h. Most of the bacteria of major importance in raw milk will grow under these conditions. However, viruses, products of microbial metabolism and somatic cells are not detected, and no differentiation of any potentially pathogenic organisms is achieved. Some have suggested that the plate count is a better index of sanitary conditions on the farm if the sample is incubated at 12·8 °C (55 °F) for 18 h before plating. The reasoning is that utensil contaminants will grow at this temperature, while the flora of the udder will not. Standards of 100 000 or 200 000 $ml^{-1}$ following pre-incubation have been suggested.

Direct microscopic counts and dye reduction tests are used in a number of areas, but they are not looked upon with favour when refrigeration and sanitation have become adequate. While the direct microscopic count does give results quickly and permits detection of excessive somatic cells, it is inapplicable to low count milk, because the individual field examined represents only about 0·000 002 ml of milk. In addition, some types of psychrotrophic and thermoduric bacteria stain poorly. The dye reduction tests are inadequate both because many thermoduric and psychrotrophic organisms reduce the dyes weakly, and because of poor suitability for low count milk. In areas where room exists for considerable improvement of poor milk supply, these tests unquestionably have some value in detecting those suppliers whose product is of the poorest quality.

Organoleptically detectable levels of change commonly involve bacterial populations in excess of $10^6$ $ml^{-1}$, and frequently $10^7$ $ml^{-1}$. Occasionally, even populations of $10^8$ $ml^{-1}$ or more, will not cause an organoleptically-detectable change, because of the relatively low level of relevant biochemical activity. Since the numbers of micro-organisms necessary to produce organoleptic changes are considerably in excess of those permitted in countries with advanced dairy technology for milk to be processed as fluid milk, such changes are not usually a problem under these circumstances. Where production and handling conditions permit excessive contamination and/or growth of micro-organisms in producer milk, defects attributable to micro-organisms become a major problem in raw milk supplies.

With more adequate farm cooling being associated with longer holding of raw milk on the farm, growth of psychrotrophic bacteria (those which can grow at refrigeration temperatures) has assumed increased importance. These bacteria are rarely present in the bovine udder, but they are widely distributed in the environment in which milk is produced and handled. Improperly cleaned and inadequately microbiocidally treated equipment is a major source, but water, dust, soil and vegetable materials are other common sources, although usually of lesser importance. Water supplies which are satisfactory from the public health standpoint (essentially free from coliform bacteria) may contain considerable numbers of psychrotrophic bacteria, and equipment rinsed with such water may be a significant source of these organisms. Even under good sanitary conditions, some psychrophilic bacteria may be expected to get into the milk. An actively growing culture of some members of this group can be expected to double in population in 4–5 h at 5 °C (41 °F), so considerable care must be exercised to limit contamination and restrict holding time to minimise populations of these organisms in raw milk.

Another aspect of the microbiological quality of raw milk is the increasing awareness that extensive growth of micro-organisms may produce enzymes which are resistant to pasteurisation (and sometimes to 'sterilisation'), and that this may permit changes after the micro-organisms responsible for the enzyme production have been killed (Law, 1979). Alterations in both flavour and physical characteristics may occur. The exact populations of micro-organisms necessary to produce enzyme levels causing these changes will depend on many factors, so that exact statements of numbers necessary for any particular change to reach a detectable level cannot be made. However, one can be reasonably sure that if the population does not exceed $10^5\,ml^{-1}$, the enzyme level will not be troublesome.

The raw milk supply may contain micro-organisms that can cause human illness, but routine tests are not made for these. Pasteurisation or an equivalent treatment will kill these micro-organisms. However, their presence in the incoming milk could be the means of contaminating the plant environment and, thus, serve as a potential source of a most undesirable form of post-pasteurisation contamination. Some staphylococci produce enterotoxins which are stable to pasteurisation, so growth of these organisms in the raw milk must be prevented, usually by holding the milk below 10 °C (50 °F).

Thermoduric bacteria, those which survive normal pasteurisation, must be at a low level in the raw milk supply if the pasteurised product is to satisfy the usual standards. These bacteria have gained considerable acceptance as an index of equipment sanitation on the producing farm, and laboratory pasteurisation at 62·8 °C (145 °F) for 30 min, followed by a standard plate count, is the usual procedure for the detection of these organisms in the milk of individual producers.

Tests for abnormal milk, primarily due to mastitis, are used extensively, and Gordon *et al.* (1980) have prepared a review covering methods of detecting abnormal milk. Although the methodology chosen may vary from area to area, either the count of somatic cells, or a test which correlates well with this count, is usually employed. The various screening and confirmatory tests are described in the standard methods (APHA, 1978), which suggest a standard equivalent to a somatic cell count of 500 000 $ml^{-1}$, or less; standards ranging from $<300\,000$ to 1 500 000 $ml^{-1}$ are being used in various jurisdictions. In homogenised milk, where the somatic cells are not carried into the cream layer as the fat globules rise, they may form an undesirable layer on the bottom of the container.

The incoming milk should also be free of antibiotics. These compounds are used frequently, sometimes almost indiscriminately, in the treatment of udder infections. The antibiotic preparation customarily carries a warning against use of the milk for human food for a period of 72–96 h after administration. Various tests, which depend upon the inhibition of a sensitive organism, have been used to detect antibiotics in milk, and antibiotics are unquestionably foreign substances in milk. However, the clinical or experimental evidence that antibiotic residues in milk have induced antibiotic sensitivity in a consumer, have resulted in anaphylactic shock when consumed by a sensitive person, or have been responsible for the development of antibiotic-resistant strains of micro-organisms of potential medical importance, is lacking.

Testing for pesticide residues, mycotoxins and other non-normal compounds which could be potential carcinogens, or which may have other potentially deleterious effects upon the consumer, are commonly of such nature that they should be carried out in laboratories that have properly trained personnel and the specialised equipment. Mycotoxins, particularly aflatoxins, may be present in the milk because of consumption by the cow of feeds on which mycotoxigenic moulds have grown. The aflatoxins found in milk are predominantly $M_1$ and $M_2$, forms less toxic than the $B_1$, $B_2$, $G_1$ and $G_2$ forms which are the major ones produced by *Aspergillus* and *Penicillium* growing on the feeds. Approximately 1 % of the $B_1$ ingested appears as $M_1$ in the milk, and $M_1$ has about 3 % of the mutagenicity of $B_1$ (Stoloff, 1980). The treatment given to milk during processing does not remove or inactivate aflatoxins. The US Food and Drug Administration considers milk containing 0·000 5 or more parts of aflatoxin per million parts of milk to be 'adulterated with harmful substance'. This figure, apparently, is based upon what can be accomplished by following good technological practices, particularly with respect to levels of aflatoxin in the peanuts and cottonseed which have been implicated as the feeds responsible, rather than on any clinical or experimental evidence that concentrations in excess of this amount will cause human illness.

## MILK PROCESSING

In the usual commercial processing of market milk, its microbiological characteristics may be influenced by storage of the raw milk prior to processing, and by pumping, filtration, clarification, standardisation, pasteurisation or sterilisation, contamination (primarily from equipment, containers and air), cooling, storage and distribution practices.

### Miscellaneous Operations

Producer milk may be held for 2 or 3 days, or even longer, on the farms when bulk tanks with presumably adequate refrigeration (7·2 °C or 45 °F) are used. The raw milk may be processed almost immediately upon receipt at the plant but, under some circumstances, it may be held for as long as 48 h. This holding period will permit some growth of the psychrotrophic bacteria almost invariably present. In addition, contamination may occur from the plant environment, particularly the equipment with which the milk comes in contact. The pasteurised milk ordinance provides that unprocessed milk shall be held below 10 °C (50 °F), but good manufacturing practice would reduce this to 5 °C (41 °F), or less, if holding is to be for more than a few hours.

Pumping of milk may result in an increase in the bacterial count of the product, even when the equipment is sterile. Agitation may break up clumps of bacteria, resulting in an increase in count without an increase in actual cell numbers. The same situation may arise when milk is agitated in a storage vat, usually to disperse the fat globules uniformly. Pumps have been responsible for significant contamination because of poor construction, or inadequate cleaning and microbiocidal treatments which have permitted microbial populations to build up and contaminate the product being pumped. Inadequate packing or poor drainage, which allows the build up of residual moisture in which micro-organisms may grow, are two of the most common problems.

Filtration has been replaced by clarification in many of the modern plants, but filtration will only remove the 'macroparticles' suspended in milk. Since micro-organisms are frequently associated with particulate material, some of the micro-organisms will be held on the filter. Ultrafiltration, through filters such as sintered glass or microporous cellulose derivatives, is impractical for milk, for a filter fine enough to be effective microbiologically will clog quickly because fat globules and other suspended materials will block the pores.

The centrifugal forces involved in clarification, separation and mechanical standardisation will throw some of the bacteria into the sludge which accumulates in the bowl of the centrifugal device used. The sludge also contains other insoluble materials such as somatic cells and coagulated proteins, as well as the type of material known as 'sediment'. Although the sludge has high bacterial counts per gram, the amount of sludge is so small in comparison with the volume of milk treated, and the break up of clumps by agitation attendant with the centrifugation is so marked, that the count in the milk is seldom reduced appreciably. Slowing the milk flow does increase microbial removal to some degree, probably simply because a given portion of the product is subjected to centrifugal force for a greater period of time. Homogenisation tends to increase this removal of bacteria, presumably because the markedly smaller fat globules do not have the same 'sweeping' effect possessed by the larger globules and their aggregates in the unhomogenised milk.

Several investigators have attempted to achieve greater microbial removal by using greatly increased centrifugal force. Some success has resulted, but potential pathogens and some spoilage organisms are not removed totally, so the process cannot serve as a substitute for pasteurisation, or any other process of similar microbiological effect.

Where standardisation of either fat or milk-solids-not-fat is carried out by mixing with such materials as skim-milk, dried milk or concentrated milk, the products used to achieve the modifications in composition must be of a quality at least equal to that of the basic product. In addition to being derived from milk of equal grade, the processing is specified to be done in such a way and under such conditions as to maintain the initial quality. The Food and Drug Administration has developed an ordinance to cover this situation (USFDA, 1978*b*).

Homogenisation of the milk, i.e. the treatment to reduce the size of the fat globules to a degree that will permit holding of the product without any significant formation of a region of concentration of milk fat (a 'cream line' in a non-homogenised product), may have two significant microbiological effects. First, an operation that breaks up fat globules also tends to break up aggregates of bacterial cells, thus increasing the colony count. While secondly, considerable care must be taken to avoid microbial contamination from the equipment used for homogenisation; improved design has made cleaning easier. However, great care must be exercised to ensure that the machine is kept in top mechanical condition, so that it can be clean and sanitary. From the microbiological standpoint, homogenisation optimally should take place after the product has been preheated almost to the pasteurisation temperature. Milk lipase will have been inactivated, and the full microbiocidal effect of the final heat treatment will be obtained after the chance for contamination from the homogeniser and related equipment has passed.

**Pasteurisation and Sterilisation**

The Grade A pasteurised milk ordinance and numerous other ordinances or regulations recognise only the use of heat for the destruction of micro-organisms in milk that is to be processed for human consumption as fluid milk. The term 'pasteurisation' is widely used for the heat treatment designed to kill pathogenic micro-organisms, but not produce a sterile product. Other forms of treatment to remove or kill micro-organisms in milk have been suggested from time to time. Some have been used commercially, but usually in conjunction with heat, or on products other than market milk. The addition of hydrogen peroxide, followed by removal of excess with the enzyme catalase, has a definite microbiocidal effect under proper conditions. Ultra-violet light can be microbiocidal when used at a sufficiently high intensity and on a very thin film of the product. Ultra-high speed centrifugation, the passage of an electric current and the addition of selective antibiotics, all have had their proponents. None of these procedures, in a commercially applicable form, has passed the test of completely eliminating potential causative agents of human disease while allowing the milk to retain its normal flavour and other desirable characteristics.

Pasteurisation of milk carried out properly accomplishes several desirable objectives. The major public health objective is the destruction of agents of infectious disease which might be present. The minimum exposures permitted have been determined by what is required to kill the most heat resistant of these agents, the rickettsia *Coxiella burnetii*. This objective alone would justify widespread use of the process.

The increase in keeping quality, particularly when refrigeration at 7·2 °C (45 °F) or less is used, is an extremely important advantage of pasteurisation. The Gram-negative psychrotrophic bacteria responsible for most defects of refrigerated milk are killed by pasteurisation. The Gram-positive cocci and rods, which survive pasteurisation to some degree, grow very slowly under good refrigeration, and ordinarily will be a cause of defects only after prolonged holding under refrigeration or when the holding temperature has been considerably higher than desirable. When post-pasteurisation contamination can be avoided, pasteurisation markedly extends the time before defects appear.

Pasteurisation inactivates some enzymes, such as phosphatase and lipase, which occur naturally in milk. Natural milk lipase is inactivated at approximately 47 °C (117 °F), well below pasteurisation levels and without pasteurisation, homogenisation of milk would be impractical because of the high degree of lipolysis which would result. If lipase were not inactivated, the greatly increased ratio of surface to volume of the fat globules which results from homogenisation would so increase the rate of lipolysis that the resulting concentrations of lower fatty acids would impart a distinctly objectionable off-flavour to the product. Even with pasteurisation, great care must be taken to ensure that homogenised product is not diverted back for mixture with raw product in which lipase activity remains high.

Under some circumstances, pasteurisation may improve the flavour of the product by masking an off-flavour with heat-induced flavour components.

Some will claim that the heat treatment of milk reduces its nutritional value by unfavourably modifying certain constituents. Some feel so strongly about the processing of a 'natural' product that they will assume the risk of contracting a milk-borne disease that would have been avoided by pasteurisation. Studies carried out over a period of years have shown that children are as well nourished by pasteurised milk as by raw milk, and with less risk of disease. People trained in public health are almost unanimously in favour of pasteurisation. In many areas, pasteurisation is required for all milk that is offered for sale. Even where an expensive programme of repeated testing of the producing animals and the people involved in the handling of milk is carried out; freedom from disease on one day does not assure continued freedom on the following day(s).

Because heat treatment does reduce the cream line (without removing or destroying any of the milk fat), and because it does modify the flavour if heating is significantly above the accepted minimum levels, much effort has been put into developing pasteurisation processes which will be microbiologically effective, but have minimal effects on cream line and flavour. As increasing amounts of fluid milk are homogenised, thus avoiding problems with the cream line, and as procedures for minimising flavour changes have been developed, pasteurisation temperatures have tended to rise.

The following combinations of time and temperature are recognised as equivalent in terms of microbiocidal efficiency (USFDA, 1978*a*): 145 °F (63 °C) for 30 min, 161 °F (72 °C) for 15 s, 191 °F (89 °C) for 1 s, 194 °F (90 °C) for 0·5 s, 201 °F (94 °C) for 0·1 s, 204 °F (96 °C) for 0·05 s and 212 °F (100 °C) for 0·01 s. All portions of the product must be heated to the designated temperature, and be held for the designated period of time, to ensure that all potential agents of infectious disease will be killed. Use of higher temperatures and/or longer periods of time is common to provide greater microbiocidal activity, as well as to provide other technological advantages under some circumstances. When additional solids, with their associated protective effect, are present, as in chocolate milk, egg-nog, cream or ice cream mix, higher temperatures are required for an equivalent microbiocidal activity.

Although equipment design is extremely important for efficient operation and satisfactory cleaning, this is particularly important for pasteurisation equipment. The requirements of the Grade A pasteurised milk ordinance and of the 3A standards, promulgated jointly by the International Association of Milk, Food and Environmental Sanitarians, the United States Public Health Service and the Dairy Industry Committee are very rigid for pasteurisation equipment, and serve as an example of what a co-operative effort between interested groups can accomplish.

The use of very brief exposures of milk to temperatures above those used for ordinary pasteurisation has been of increasing interest in recent years. Mehta (1980) has reviewed much of the recent literature. The Grade A pasteurised milk ordinance defines ultra-pasteurised milk as that thermally processed 'at or above 138 °C (280 °F) for at least 2 s, either before or after packaging, so as to produce a product which has an extended shelf-life under refrigerated conditions'. Such milk will, ordinarily, be sterile at the time it leaves the heating portion of the equipment. Heating outside the final container can be accomplished by direct exposure to superheated steam, in which case the water added as steam is subsequently removed by evaporation, when the hot product is injected into a vacuum chamber designed for this purpose. Indirect heating by means of plates or tubes which separate the milk and the heating medium may also be used. What may be achieved is 'commercial sterility', i.e. some of the product is not completely without surviving micro-organisms, but those that survive are unable to grow under the conditions of storage. They could be obligate thermophiles, and thus be unable to initiate growth at ordinary storage temperatures. They could be obligate aerobes, and thus be unable to grow in the anaerobic environment of a full and hermetically sealed container. They may have been sub-lethally stressed by the heating, and thus be unable to develop in the unfavourable environment of the packaged product, but able to grow in the culture media and conditions provided by the laboratory. This product can be placed aseptically in a final package. Alternatively, it may be placed in a final package non-aseptically, and then be subjected to an additional heat treatment of sufficient intensity to kill the occasional contaminant. By another approach, the entire heating process can be carried out following the introduction of a prewarmed product into sealed containers, usually glass but, because of the slower rate of heat transfer under these conditions, the higher heat treatments at correspondingly shorter exposures cannot be used. The product has a somewhat darker colour and a more 'cooked' flavour as a result. Under favourable circumstances, products of considerable keeping quality at room temperatures, without excessive levels of flavour or colour deterioration due to the high temperatures employed, can be produced. These products obviously have considerable appeal under conditions where good refrigeration is not available, is too expensive or is undependable.

**Cooling**

Immediately following pasteurisation or sterilisation, a regenerative system is commonly used for a portion of the cooling. This is followed by a further cooling to the desired final temperature, using a refrigerated non-milk coolant. Protection against contamination is provided not only by careful design and construction of leak-proof equipment, but also by maintaining the pasteurised product at a higher pressure than the coolant. Care in maintaining gaskets and seals, and appropriate pressures on the presses, is particularly important with plate heat exchangers. Although the Grade A pasteurised milk ordinance requires the cooling of pasteurised milk only to 7·2 °C (45 °F), this should be regarded as a minimum standard. Cooling to at least 4·5 °C (40 °F), and preferably lower, is desirable because of the warm-up which may occur during subsequent handling and packaging, and also because outgrowth of surviving micro-organisms, and possible contaminating micro-organisms, will be more retarded as holding temperatures are reduced. Air cooling of the product in the final package can be very slow, both because of the poor cooling efficiency of air, and because the shape of many of the packages permits such close packing of units that air circulation between individual final package units is almost completely prevented.

**Packaging**

The equipment for filling milk into the final package is of such complexity (Fig. 1) that extreme care must be taken to prevent microbial contamination from this source. Proper drip deflectors must be installed to prevent condensate from cold surfaces getting into the final container and providing a potential source of contamination. The measuring devices, valves and other parts require very careful cleaning, storage between uses, and pre-use microbiocidal treatment to prevent this equipment from being a serious source of contamination.

FIG. 1. This modern high-speed filling unit illustrates just one of the complex components of a bottling-line for market milk, and yet inspite of the obvious potential sources of contamination, careful cleaning and sanitising ensures that the hygienic quality of the product is consistently high. Reproduced by courtesy of Unigate Foods Ltd, St. Erth.

Glass was essentially the only material used for the containers in which milk was delivered to the final consumer some years ago, but numerous other types of container materials are now used. Paper containers have become widely accepted, with plastic coating having replaced the hot paraffin coating common in former years. Rigid plastic containers may be used, particularly for the larger sizes, and non-rigid plastic containers, such as the 'pillow packs', are found in a number of markets. For each of these types, the form of closure is an important consideration, both technologically and microbiologically. Some containers are single use, while others are returned for re-use, after proper cleaning and microbiocidal treatment. All forms of containers must be made and sealed with non-toxic materials, that is materials which will not render the milk injurious to health, or which may adversely affect the flavour, odour or microbiological quality of the product, and which otherwise meet the requirements of the regulatory agencies.

To be economically competitive in the milk industry, the glass container must be re-used repeatedly. A glass container undoubtedly is sterile immediately after being blown from the hot molten glass, but it is exposed to microbial contamination as it is handled. All glass containers must be subjected to cleaning and microbiocidal treatments immediately prior to use, and must be adequately protected against contamination as they pass from treatment area to filling area. The concentrated alkali solution used at elevated temperatures as the detergent in most mechanical bottle washing procedures has a marked microbiocidal effect. The final chlorinated water rinse must be maintained at a concentration which not only destroys those micro-organisms which may be in the untreated water, but also those which may have survived the cleaning and preliminary rinse cycles of the washer. Hot water or steam may also be used for the final microbiocidal treatment. The usual standard applied is that the container shall contribute less than one micro-organism per ml of capacity, and no coliform bacteria. Of course, whatever residual micro-organisms there are should neither be capable of causing human illness, nor be psychrotrophic and, thus, potential contributors to poor keeping quality of product.

Paper containers have replaced glass in many markets, at least in part, because they are single-use containers that are advantageous for marketing through the increasingly important retail outlets. The problems of holding for return, and of monetary deposits to promote return, do not exist with paper. In addition, the lighter weight of the container per unit volume is a distinct advantage. Specifications for processing the pulp, and producing the final container have been developed to minimise the microbial content of the finished carton (USFDA, 1978*c*). Among the manufacturing steps which have a microbiocidal effect are the bleaching of the pulp with chlorine, the sizing (during which a low pH is attained) and the drying process (during which temperatures in the range of 121 °C (250 °F) are reached). The processes employed quite consistently result in containers meeting microbiological standards, but protection of the blanks from contamination during storage, prior to use, is essential.

The sheet used for the non-rigid plastic container is produced under conditions that raise temperatures to the point where micro-organisms do not survive. The microbiological concern is with the protection from subsequent contamination during handling, forming into the final package, and operations associated with the filling and closure. The rigid plastic container may be blow-moulded either at the point of use or, more commonly, at a central unit and transported to the point of final use. The temperatures employed during 'melt' of the plastic and the moulding are highly microbiocidal. The microbiological consideration, again, is protection against subsequent contamination.

The containers used for aseptic packaging of previously sterilised products must undergo a microbiocidal treatment that can be depended upon to give 'commercial sterilisation'. Where cans are used, they are usually heated by gas flames to a level which will destroy any micro-organisms present. Both glass and metal containers can be subjected to steam under pressure as the microbiocidal agent. Total heat treatment would be destructive to the containers laminated from paper, plastic and in some cases, a metal foil; a low level of initial contamination and a chemical treatment are usually relied upon. Hydrogen peroxide treatment followed by a level of heating that will remove the residual chemical is used in some procedures.

The complexity of the equipment used to provide closure, whether by applying a cap or by sealing with a combination of heat, pressure and/or a sealant, presents some microbiological problems. Proper design, adequate cleaning and a microbiocidal treatment are essential. Caps, hoods and other materials must be protected from contamination during storage and handling. Capping by hand should never be permitted, as the person involved may be a source of contamination. Overfill must be avoided to prevent product from coming out through or around the closure, and thus possibly being sucked back when the condition causing the outflow is reversed, as this could introduce deleterious micro-organisms into the packaged product.

Particularly when 'sterilised' milk packages are being handled, rough handling must be avoided, as this may result in small or even temporary breaks in the package or closure. Even a momentary small opening may permit micro-organisms to gain entry to the product and increase the potential for spoilage. This has been exhaustively documented in the canning industry, but it also applies to other types of containers. The presence of moisture on the container surfaces increases the chance of contamination, the contaminated moisture being more apt to penetrate the container break than would dry air.

## Distribution

The handling of milk after it leaves the refrigerated storage of the processor can greatly influence its microbiological condition. The trucks in which milk is transported to either the store from which it is sold, or the home, or other outlet in which it is consumed, should have facilities adequate for maintaining the temperature below 7·2 °C (45 °F). Under winter conditions in areas of cold weather, nature may provide the necessary cooling, and provision frequently must be made to avoid freezing. When atmospheric temperatures are higher, either mechanical refrigeration or adequate amounts of ice, properly distributed, are needed. At the unloading dock, whether at a store, restaurant or other facility, the product should never be left for more than a very few minutes before being moved to an area of adequate refrigeration. Some surveys of holding rooms in stores have shown temperatures well above the desired maximum of 7·2 °C (45 °F). In the case of home delivery, an insulated, light-proof box should be used for milk left on the doorstep; not only will direct sunlight rapidly increase temperature, but it may also cause light-induced off-flavours. Protection from contamination by cats, dogs, birds and other potential forms of tampering is desirable.

The manner in which milk is offered for sale in the retail outlet is important. Cabinets with adequate closure and proper provision for maintaining uniform temperature, despite frequent access, are essential. Open-top display cabinets must be extremely carefully designed and operated; they are not permitted in some areas because of the problems associated with their use. Thermometers in the critical areas of all display cases should be read at frequent intervals. Temperatures of milk in representative containers should be determined periodically, particularly when any evidence of malfunction of the cooling system exists. Rotation of stock is exceedingly important and should, now that most product is code-dated, involve few problems beyond adequate supervision. Since almost all lots of pasteurised milk have a finite keeping quality, even under good refrigeration, failure to merchandise the oldest product first is an invitation to consumer problems.

In the home, milk should be refrigerated at around 7·2 °C (45 °F) (preferably somewhat colder), and now that milk will be held for as long as a week and sometimes longer in the home, care must be taken to avoid any periods of warm-ups, such as by keeping a container outside the refrigerator for more than a few minutes. Product removed from the package and unused should never be returned to the package, because of the almost certainty that it will have become contaminated with potential spoilage organisms. In most homes, restaurants and similar situations, the microbiocidal treatment of utensils is inadequate for destruction of all micro-organisms, and subsequent contamination from dish towels, hands, dust, air and other potential sources is almost unavoidable.

Temperature increases have serveral microbiologically undesirable effects. Higher temperatures will increase the rate of growth of organisms which were growing, probably very slowly, at the lower temperature; Table I shows how this affects several strains of bacteria isolated from milk. As

TABLE I

GENERATION TIMES OF REPRESENTATIVE PSYCHROTROPHIC BACTERIA ISOLATED FROM MILK AND GROWN IN STERILE MILK AT VARIOUS TEMPERATURES[a]

| *Culture* | *Generation time (min) at temperature of:* | | | | |
|---|---|---|---|---|---|
| | *5°C* | *10°C* | *21°C* | *25°C* | *32°C* |
| *Pseudomonas ovalis* | 255 | 255 | 83 | 90 | 100 |
| *Pseudomonas arvilla* | 231 | 217 | 96 | 108 | 188 |
| *Pseudomonas geniculata* | 217 | 175 | 108 | 83 | 104 |
| *Pseudomonas* sp. | 222 | 221 | 108 | 82 | 54 |
| *Pseudomonas fragi* | 231 | 199 | 80 | 86 | 207 |
| *Flavobacterium aquatile* | 285 | 280 | 433 | 433 | No growth |

[a] Adapted from data of Lawton and Nelson (1954).

temperature rises, some types of bacteria which could not grow at the lower temperatures will be able to grow. Some organisms, sub-lethally injured by factors such as heat and cold, are more able to recover from injury and initiate growth as temperatures rise from the desired refrigeration range. Once the temperature has been allowed to rise, recooling may be quite slow, particularly when air is the cooling medium in contact with the container.

**The Cleaning and Microbiocidal Treatment of Equipment**

The basic principles are the same as for equipment on the farm, except that heating of milk frequently results in residues which adhere tenaciously to the heating surfaces. These heating surfaces require special care to avoid the build-up of milk deposits. Hand-brushing of dairy equipment has largely been replaced by cleaning-in-place (CIP) procedures, both on the farm and in the processing plant, and this change involves designing the system specifically for this procedure. In the early use of CIP, a significant amount of disconnecting of some lines, capping of others, removal of some equipment from the circuit for hand cleaning and other manual activity was required. More recently much of this muddle has been eliminated, so the majority of the equipment can be cleaned in place, and the modifications in flow necessary to change from processing to cleaning are brought about by a series of interconnected, remotely actuated valves. The object has been to reduce human participation, with the attendant chance for human error, to the minimum. A number of producers of equipment, detergents and microbiocidal materials, as well as some industrial consultants, are available to provide appropriate assistance in the development and operation of CIP systems.

In all cleaning procedures, several basic steps are involved. First, following draining-off of the product as completely as possible, the major residues must be rinsed from the equipment, preferably using water at 37·8–49 °C (100–120 °F). This is warm enough to dissolve soluble residues and to liquefy food fats, but not so hot as to 'bake' the residues onto the product-contact surfaces. Secondly, a detergent solution is employed to remove the residual soil, with appropriate force being supplied by brushing, high pressure jets or rapid circulation. When hand cleaning is employed, both the temperature of the solution and the kind and concentration of the detergent must be kept at levels tolerable to the human hand; with CIP procedures these limitations are avoided. The type of detergent varies considerably with the procedure being employed and the characteristics of the water used. Alkaline agents are most common, but acids are often used from time to time to remove soil which may build up with the continued use of alkali. The exact procedure should be worked out with specialists who are knowledgeable about the particular situation. Thirdly, the equipment is drained, and then flushed with warm water to remove the detergent solution and the soil suspended and dissolved therein. Fourthly, the equipment is given an appropriate microbiocidal treatment, preferably just before use. Steam, hot water or chemical agents, such as chlorine, may be used, but one must remember that chemical agents are microbiocidal only when they actually contact, at adequate concentration, the organism against which they are to act. Residual soil may not only protect the organism from direct contact with the chemical agent, but also dissipate some of the microbiocidal activity by chemical reaction. The equipment should be kept dry during storage, as residual moisture may permit surviving micro-organisms to grow. Properly carried out, procedures based on these essential steps will leave very few micro-organisms, and those remaining will be of little practical importance in properly refrigerated milk. Where absolute sterility is necessary, as in aseptic packaging areas, special procedures, frequently involving steam under pressure, must be used, as not even low levels of mesophilic spore-forming bacteria can be tolerated in these circumstances.

CIP cleaning has the advantages of low labour cost, less damage to lines and equipment (through reduced handling), usually less product loss (because of better drainage and less damage to lines and fittings), and more consistent results arising from automation, and because higher solution concentrations and temperatures can be employed. Provisions must be made for adequate monitoring of temperatures, concentrations and final cleanliness of equipment. For the latter, resort must be made to inspection ports, special sampling ports, partial disassembly and appropriate laboratory tests for residual micro-organisms. One disadvantage of CIP is the considerable cost of the equipment and controls that are required for satisfactory operation.

For a CIP system to be fully effective, certain precautions must be taken. The lines must be firmly supported, with fittings and gaskets in alignment, so that leakage between gaskets does not result in ineffective cleaning; gaskets should form an essentially flush joint. All lines and equipment must drain properly, and lines should be pitched at about 1 in. in 20 ft. A recording thermometer in the return line should provide a dependable record of time and temperature for each CIP operation. A solution tank (or tanks) of appropriate size, and a pump (or pumps) of adequate capacity to completely fill the lines and provide adequate solution velocity must be provided. The solution tanks should be heated by a direct steam line with a thermostatic control to ensure proper temperatures. The detergent(s) must be chosen for suitability for the particular system, and must be used at the appropriate concentration(s). The microbiocidal agent(s) employed, whether depending upon heat or chemical action, must be used at appropriate levels and for adequate periods of time.

## MICROBIOLOGICAL PROBLEMS WITH MARKET MILK

Concern with micro-organisms in pasteurised market milk and related products involves primarily the bacteria. Yeasts and moulds ordinarily constitute no problem, mainly because the types encountered both do not survive normal pasteurisation, and the chance contaminant does not grow under the conditions of storage employed. Viruses and rickettsia must be considered in relation to the possible transmission of disease. They are not problems in terms of spoilage, and procedures for their detection are not applicable to routine control laboratories. Procedures for the detection and enumeration of the relevant bacteria are outlined in a number of publications, the one most applicable to the products under consideration being *Standard Methods for Examination of Dairy Products*, the 14th edition of which appeared in 1978 (APHA, 1978).

**'Total' Counts on Pasteurised Milk**

The limit of 20 000 cfu $ml^{-1}$ for the plate count on pasteurised milk, to which no culture organisms have been added, is the result of years of experience showing that this is achievable, usually with some margin, when adequate attention is given to the quality of the incoming milk, and to good manufacturing practices. Over a period of years, the number has been decreased somewhat from previous standards. However, the real change has been that the use of improved culture media and lower incubation temperatures has resulted in an increased efficiency in enumeration of the bacteria which survive pasteurisation. The beef extract–peptone medium used in the early years did not support the growth of many types of bacteria important in dairy products. Later studies have shown that the medium also failed to permit colony production by many cells that had been sub-lethally stressed, as by pasteurisation. The early use of the medically dictated 37 °C (98·6 °F) also resulted in an under-enumeration of bacteria, and even 32 °C, as now used occasionally, results in missing some bacteria, such as certain types of psychrotrophic organisms found in products held for some time under good refrigeration. Extending incubation times beyond $48 \pm 3$ h, the substitution of certain other peptones for the tryptic digest of casein, and modifications in pH of the medium have been found to result in higher counts on some pasteurised products. No one combination of conditions for the plate count can be expected to permit colony formation for all of the bacterial types in their various physiological states found in dairy products. Since variables do influence counts, standardisation of procedures is essential if results of tests made in different laboratories are to be comparable.

The standard plate count on pasteurised milk and related products is something of an index of good manufacturing practices. If the raw product was of satisfactory microbiological quality, the processing was carried out properly, protection from contamination was satisfactory, holding temperatures and times were such that growth of bacteria did not occur to any significant extent, and other less important factors were kept under proper control, the count will be low. However, a low count does not ensure that the product will be free from potential disease-producing organisms. A negative phosphatase test, indicating satisfactory pasteurisation, is a much more useful criterion. A low count does not assure freedom from contamination with bacteria which will cause poor keeping quality and, as will be discussed in relation to psychrotrophic organisms, the contamination level that may cause considerable problems is usually well below the level that will result in a significant change in the plate count.

The other microbiological tests for 'total microbial populations' that are sometimes applied to pasteurised products are of strictly limited applicability, and frequently can be criticised as leading to a false sense of security. The populations should be so low that the dye reduction tests are inapplicable, no matter what dye, or what pre-incubation procedure, is used. The long incubation times required for dye reduction with such low count milk are more a measure of growth than of initial population. Both thermoduric organisms, which have survived pasteurisation, and psychrotrophic organisms, which may have grown following contamination, are notably poor reducers of the dyes. Organisms, such as *Streptococcus lactis* and the coliform bacteria, that are good reducers of the dyes should not be present, being unable to survive pasteurisation. Preliminary incubation before the dye reduction test will give a result somewhat indicative of the growth rate of the organisms at the preliminary incubation time and temperature chosen, rather than an index of the bacterial population of the initial sample.

A direct microscopic count on pasteurised milk will tell if the population has reached a level well above what is usually acceptable, but with some definite limitations. Differentiation of living and dead cells is usually difficult. The living cells, even if detectable, should be at a level well below the applicability of the test, since each field examined represents only about 0·000 002 ml of the product. If one had reason to be concerned about thermophilic bacteria, or the occasional case where Gram-positive psychrotrophic bacteria might have grown extensively, both the numbers present and the ready stainability of these types make the direct microscopic procedure applicable. As some of the tests for biomass are refined and further developed, they could offer promise for the estimation of microbial populations in products such as pasteurised milk. At the present level of knowledge, the standard plate count, or some of the modifications which have been introduced to reduce amount of equipment and time required in routine control, remain the procedures of choice for the estimation of the total microbial population in pasteurised milk and related products.

No microbiological test can serve as adequately as the phosphatase test in determining whether pasteurisation, and protection from post-pasteurisation contamination with unpasteurised product, has been carried out adequately. Milk phosphatase has the fortunate characteristic of being inactivated at heat exposures just less than pasteurisation. Some reactivation of the enzyme may occur during storage after high temperature–short time pasteurisation, particularly with products of increased fat content. Some micro-organisms produce enzymes which may be confused with natural milk phosphatase, but adequate procedures have been developed to differentiate those enzymes which might otherwise give misleading results. (APHA, 1978).

**Thermoduric Bacteria**

The organisms that survive pasteurisation, but do not grow at pasteurisation temperatures, are considered by the dairy industry to be thermoduric. The degree of survival can range all the way from a fraction of 1 % of the original population to an actual increase in count, as in the case of refrigerated cultures of *Microbacterium lacticum*; this latter situation possibly results from the breakup of cell aggregates by the heating. Thermophilic bacteria are excluded by this definition because they grow at least at the lower range of pasteurisation temperatures. Thermoduric organisms are predominantly mesophilic, but a few types are included among the psychrotrophic organisms to be discussed later, since they grow slowly at refrigeration temperatures.

Thermoduric populations are usually determined by laboratory pasteurisation of the sample, followed by a plate count to determine the surviving population. A combination of 62·8 °C (145 °F) and 30 min holding, after the requisite temperature has been reached, is ordinarily used. The difficult mechanics of using high temperature–short time pasteurisation procedures for routine laboratory examination of small samples have largely precluded the use of this procedure. The surviving cells have been sublethally stressed, and are, thus, more demanding than unstressed cells in the conditions necessary for colony formation. Factors, such as time and temperature of plate incubation, pH, and type of peptone in the plating medium, can all influence the counts of thermoduric micro-organisms, and the distribution pattern between genera can also be modified appreciably by the plating conditions.

The commonly encountered thermoduric bacteria are found in the genera *Arthrobacter*, *Bacillus*, *Microbacterium*, *Micrococcus* and *Streptococcus*. *Lactobacillus* spp. are encountered less commonly, probably, in part, because of considerable problems in having the conditions of enumeration satisfactory for colony development. A number of other genera have been encountered among thermoduric populations, but so uncommonly as to be of little practical importance.

The genera *Micrococcus* and *Staphylococcus* must be differentiated to avoid the considerable confusion that has existed. Members of the genus *Micrococcus* have a strictly respiratory metabolism, oxidising glucose to acetate or to carbon dioxide and water. Members of the genus *Staphylococcus* ferment glucose anaerobically to lactic acid. The guanine plus cytosine content of DNA is 66–73 mol. % for the genus *Micrococcus*, and 30–38 mol. % for the genus *Staphylococcus*. The members of the genus *Staphylococcus* are not thermoduric, and they are mainly infectious agents associated with the animal body, including the cow's udder. Many members of the genus *Micrococcus* are thermoduric, and are frequently associated with poor sanitation of equipment used for handling milk; they are present in insignificant numbers in milk drawn aseptically. They commonly constitute the major fraction of the thermoduric population of milk.

The genus *Microbacterium* is included here (although *Bergey's Manual of Determinative Bacteriology* (Buchanan and Gibbons, 1974) has classified it among the *genera incertae sedis* associated with the genus *Arthrobacter*), and its members are Gram-positive, catalase-positive, aerobic, non-spore-forming, non-motile short rods of relatively small size; organisms of this genus can ferment lactose to give some lactic acid. *Microbacterium lacticum* is one of the most heat-resistant of all non-sporulating, mesophilic bacteria known; some strains have 11–33 % survival after heating at 84 °C for 2·5 min., and a few cells can survive for 15 min at the same temperature in milk. The genus also includes *M. liquefaciens*, which has been isolated frequently from milk in Europe, but not in the United States.

Because many strains of microbacteria fail to grow at 35 °C, and produce countable colonies only slowly at 32 °C, they have only been recognised as a probable major component of the thermoduric microflora in the last 20–25 years. If lower incubation temperatures or longer incubation times were to be used for the plate count, these bacteria would be encountered in considerable numbers.

The genus *Streptococcus* includes several species which are thermoduric, in addition to the non-thermoduric lactic streptococci (*Str. lactis*, *Str. cremoris* and *Str. lactis* sub-sp. *diacetylactis*) and the non-thermoduric streptococci which are agents of human and animal disease. The species *Str. thermophilus*, *Str. faecalis* and its varieties and *Str. faecium* are the principal thermoduric streptococci. Occasionally the thermoduric, β-hemolytic *Str. faecalis* var. *zymogenes* is found in pasteurised milk, and causes some concern until shown not to be *Str. pyogenes* or some other potential agent of human illness. Thomas *et al.* (1966*a*, *b*) have shown that substitution of a more suitable peptone for tryptone, and adjustment of the medium to pH 7·5 or above, very considerably increase the recovery of thermoduric streptococci from pasteurised milk.

Members of the genus *Lactobacillus* are not encountered very frequently among the colonies developing when thermoduric counts are made by the usual procedures, although Slatter and Halvorson (1947) have shown that some members of the genus have sufficient heat resistance to contribute significantly to thermoduric counts. Thomas *et al.* (1963) provided a possible explanation in that, while none of the colonies counted after plate incubation for 2 days at 35 or 32 °C were lactobacilli, an appreciable number of colonies which developed during 3 and 4 days of incubation were those of lactobacilli. When one considers that many lactobacilli grow poorly under aerobic conditions, and also require a rather complex medium for growth, and then combines this knowledge with the well-documented observation that sub-lethally stressed organisms are more exacting in their requirements for growth and colony formation, the failure to recover lactobacilli from many pasteurised samples, when 'usual' or 'standard' procedures are used, is understandable.

The genus *Arthrobacter* is also included here, although the correctness of this classification becomes questionable when *Bergey's Manual* (Buchanan and Gibbons, 1974) states that members of the genus *Arthrobacter* 'do not survive heating at 63 °C for 30 min in skim-milk'. However, species which were assigned to this genus have been found to constitute a significant percentage of the thermoduric population of milk. The incubation of plates at 35 °C, as was done in many of the earlier studies, reduces the number of colonies of this genus, as compared with incubation at 28 or 32 °C.

The genus *Bacillus* includes those aerobic, spore-forming bacteria that are almost invariably present in small numbers. When the thermoduric population of the milk is low, *Bacillus* spp. may make up a significant fraction of the total thermoduric population. When thermoduric populations are high, *Bacillus* spp. usually constitute only a small fraction, presumably because they do not grow as readily on the equipment as do some of the other thermoduric types. *Bacillus subtilis* and *B. cereus* are the species encountered most commonly, although a number of other species have been identified. Neither their nutritional requirements nor their temperature range for growth appear to present any problems in the enumeration of these organisms. The tendency of some strains to form spreading colonies may interfere with enumeration of other thermoduric types, and low dilutions must be employed.

Organisms of the genus *Clostridium* have been found among the thermoduric flora, but usually at a very low level. The incidence probably would be somewhat greater if anaerobic techniques were used, although the numbers encountered would probably still be small.

Occasional reports of the isolation from milk or milk products of thermoduric bacteria that do not belong to any of these groups do appear, but their exact taxonomy has not been determined in many instances.

*Escherichia coli* and *Enterobacter aerogenes* ordinarily do not survive pasteurisation. Occasional instances of survival in fairly large numbers have been reported, but some of these reports are quite probably the result of faulty laboratory techniques. Practical plant experience indicates that the coliform bacteria should not be included among thermoduric micro-organisms.

None of the bacteria known to cause human infectious disease have been shown to survive normal pasteurisation.

**Sources of Thermoduric Bacteria**

In the pasteurising plant, heat-resistant bacteria can come from improperly cleaned equipment, but this is uncommon in a well-operated plant. Repasteurisation of returned milk, which has been allowed to warm enough to permit growth of micro-organisms surviving the first pasteurisation, can result in high thermoduric counts, but the problem is usually due to high counts of thermoduric bacteria in the raw milk supply. A count of 10 000 cfu $ml^{-1}$ following laboratory pasteurisation has been proposed as the permissible upper limit for the thermoduric count. Many producers can better this standard by a substantial margin.

Thermoduric bacteria are usually considered an index of equipment sanitation on the producing farm. They survive inadequate cleaning and microbiocidal treatments, and grow in residual moisture. In the opinion of this writer, a better correlation between equipment care and thermoduric counts might well result if the enumeration conditions employed were better suited to maximum recovery of these organisms, but data on this possibility do not seem to be available. A tendency exists for thermoduric counts to be higher in summer than in winter, probably because higher environmental temperatures provide more opportunity for growth. Cooling of the raw milk has little effect on thermoduric counts, presumably because these mesophilic organisms are all either slow in growing, or are unable to grow at the temperature levels effective in slowing down spoilage by psychrotrophic bacteria.

With the exception of a few strains of enterococci and an occasional

*Bacillus* or *Clostridium* species, thermoduric bacteria are unable to grow at or below 7·2 °C (45 °F) and are, thus, not a factor of importance in the spoilage of pasteurised milk. Above this temperature they do grow, and may be a major factor in the spoilage of milk and other dairy products not held properly refrigerated. Instead of the clean, acid spoilage characteristic of *Str. lactis*, thermoduric bacteria may be expected to cause undesirable, unclean flavours, and sometimes some physical changes. In raw milk, the thermoduric types will almost invariably be overgrown by other micro-organisms which will cause defects characteristic of their metabolic activities.

Some evidence exists suggesting that HTST pasteurisation, particularly at minimum levels, is less effective than holder-type pasteurisation in reducing populations of thermoduric micro-organisms. On occasions, raw milk supplies suspected of having high populations of thermoduric bacteria have been diverted to plants employing the holder process as a temporary means of control. Today, this could be a difficult solution to implement with the almost universal use of the HTST process.

**Thermophilic Bacteria**

Thermophilic bacteria are commonly defined as those which will grow readily at 55 °C (131 °F); mesophilic bacteria will usually be markedly retarded, or unable to grow at all above 50 °C (122 °F). Some facultatively thermophilic bacteria will grow at 37 °C (98·6 °F) or lower and, thus, may show up as colonies on the standard plate count; obligately thermophilic bacteria will not grow at 37 °C. The upper limit for growth of most thermophilic bacteria is about 70 °C (158 °F), and the standard procedure for the enumeration of thermophilic bacteria is to incubate the agar plates at 55 °C. The direct microscopic procedure is useful for detecting these organisms when present in large numbers.

The thermophilic bacteria found in milk are primarily aerobic or facultatively anaerobic spore-forming rods. Prickett (1928) has described these organisms in considerable detail. *Lactobacillus thermophilus* has been placed in the genus *Bacillus* in *Bergey's Manual*, probably being closely related to, if not identical with, *B. coagulans*.

Raw milk generally contains few thermophilic bacteria, although sufficient numbers are normally present for large numbers to develop during holding at high temperatures. Soil, bedding, feeds and occasionally, water may be sources, but the udder of the cow does not harbour these organisms.

Thermophilic bacteria become a problem in pasteurised milk when portions of the milk are held for some time within the temperature range of

50–70 °C (122–158 °F). Repeated use of the vats for batch pasteurisation, without thorough cleaning between uses, has been found to permit a build-up of thermophilic bacteria in the milk and milk-film residues. Milk foam and dead-ends holding warm milk have also caused problems. When an interruption in normal scheduling occurs, delayed cooling may permit growth and, because the pasteurisation temperature of 62·8 °C (145 °F) is so suitable for growth of thermophilic bacteria, any prolonged holding in this temperature range of milk, or the equipment with which it comes in contact, must be avoided. Repasteurisation of returned milk, or other previously pasteurised product, may permit enrichment of thermophilic bacteria already present in modest numbers.

The extensive use of HTST pasteurisation has markedly reduced problems with thermophilic bacteria. Thus, in a typical operation, less than 2 min is required for the milk to pass from being cold, raw milk through to being cold (4·4 °C or 40 °F) pasteurised milk, and the holding temperature (71·7 °C or 161 °F) and holding time (16 s) are a combination not conducive to growth of these bacteria. The regenerative section is the only place in the unit where the temperature is favourable, and a cooked-on milk film in this section could permit some growth, but this is not usually a problem.

**Psychrotrophic Bacteria**

Psychrotrophic organisms, as this term applies to the dairy industry, can be defined as those organisms capable of appreciable growth at commercial refrigeration temperatures of 2–7 °C (35–45 °F), irrespective of their optimum growth temperature. The term 'psychrophilic' should be reserved for micro-organisms whose optimum growth temperature is below 20 °C (68 °F). Very few of the psychrotrophic organisms found in dairy products are classical psychrophilic organisms, since nearly all of them are mesophilic, i.e. their optimum growth temperature is in the 20–32 °C (68–90 °F) range.

The majority of the psychrotrophic bacteria encountered in dairy products are Gram-negative, non-sporulating, oxidase-positive, small rods that commonly are placed in the genera *Pseudomonas*, *Flavobacterium* and *Alcaligenes*. Many have been placed in the genus *Achromobacter*, but this genus is no longer recognised. Some have been classified in the genus *Acinetobacter*, which is separated on the basis of being catalase-negative, among other criteria. Not only does considerable confusion exist concerning the taxonomy of bacteria in this general area, but also the organisms isolated as psychrotrophic frequently have not been characterised adequately in terms of modern microbial taxonomy.

Psychrotrophic micro-organisms of the spore-forming genera *Bacillus* and *Clostridium* have aroused considerable interest in recent years, since they are also thermoduric. When competition from the more common psychrotrophic bacteria is non-existent or minimal, these spore-forming types can sometimes develop during long holding periods under what would be considered adequate refrigeration; defects caused include sweet curdling and bitterness. *Streptococcus faecalis* and its sub-species and *Str. faecium* can grow slowly in the range for psychrotrophic organisms, and these bacteria are also thermoduric. Some strains of coliform bacteria have been found growing in properly refrigerated milk. A number of yeasts and moulds found in dairy products are also included among the psychrotrophic organisms, but they are not of practical importance in pasteurised market milk. Until a few years ago, one could say that no microbial agent of human illness associated with milk consumption would grow in milk at temperatures below 7·2 °C (45 °F). Now *Yersinia enterocolitica*, an agent of enteric disturbance, has been found able to grow in this temperature range, and has been implicated in an outbreak attributed to chocolate milk.

From the standpoint of quality control of pasteurised milk and related products, psychrotrophic bacteria, particularly of the Gram-negative group, undoubtedly are the most important organisms; their importance has increased as storage times have lengthened with changes in technology and marketing conditions. These organisms can cause a variety of flavour defects, including fruity, rancid, stale, bitter and putrid. Some can cause ropiness, and others have caused colour changes. Whereas a population of $10^7\,ml^{-1}$ will result in a fairly pronounced defect with some of these organisms, others will cause no detectable defect at 10 times this population level, because of the difference in their biochemical activity.

*Psychrotrophic Bacteria in Pasteurised Milk*

Nearly all pasteurised milk and related products held under refrigeration will eventually develop one or more defects due to psychrotrophic bacteria. The rate at which a defect develops will depend upon the initial number of these organisms present, the rate at which the organisms grow at the holding temperature used, and the ability of the organisms to cause an organoleptically detectable change in product. Theoretically, only one cell of a type capable of causing a defect needs to be present in a given container, whether the capacity be half a pint or 10 gal., although in the larger container, a few more generations will be required to bring the population to the level per ml necessary to cause the defect; the result is a slightly better 'keeping quality'.

When the thermoduric psychrotrophic bacteria of the genera *Bacillus*, *Clostridium* and *Streptococcus* are responsible for a problem, the remedy is either to reduce the numbers of these bacteria in the milk supply, or to increase the intensity of the heat treatment to a level which will kill the responsible organisms; the latter being difficult under most circumstances. Refrigeration in the upper range of acceptability will increase the probability that this group will develop, but organisms of this group do tend to be overgrown by other types of psychrotrophic bacteria.

The psychrotrophic bacteria of the Gram-negative group are killed, in most instances by a considerable margin, by normal pasteurisation, and their presence in the pasteurised product is the result of post-pasteurisation contamination. Organisms of this group tend to outgrow the thermoduric psychrotrophic types, particularly when the product is held at the lower range of temperatures. Their control depends upon preventing post-pasteurisation contamination, or at least keeping the level extremely low. Since these organisms are quite easily killed by either heat or chemical microbiocidal treatments, the basic problem usually is one of being sure that no little details are overlooked. Contamination levels so low that they are not detected by the usual microbiological tests can still contribute enough organisms for keeping quality to be a problem.

The temperature at which product is held can have a great effect on the growth of psychrotrophic bacteria. At 10 °C (50 °F) and above, non-psychrotrophic thermoduric bacteria are apt to outgrow the psychrotrophic types, but at 7·2 °C (45 °F) and below, dominance of the psychrotrophic types is found. As temperatures decrease toward 0 °C (32 °F), some of the psychrotrophic organisms will be kept from growing, and those that do grow, will do so at progressively lower rates. Data on three commercial samples with different initial populations of psychrotrophic bacteria are shown in Table II. Marked increases in rates of growth occurred as temperatures increased from 2 °C (35·5 °F) to 15 °C (59 °F), and with samples *B* and *C*, the proportion of the total population that was made up of psychrotrophic bacteria was greatly reduced by holding the samples at 10 and 15 °C; sample *C* was without defect after holding for 30 days at 2 °C. The initial coliform populations were $\leq 1\ ml^{-1}$ in all three samples, but this gave no clue as to the rate at which the psychrotrophic bacteria developed. The optimum control for psychrotrophic bacteria is to have the least possible post-pasteurisation contamination, followed by holding the product at the lowest practical temperature, and certainly not above 5 °C (41 °F). Even relatively short periods of warm-up may so accelerate growth that keeping quality is reduced significantly.

TABLE II

INFLUENCE OF HOLDING TIME AND TEMPERATURE ON COUNTS OF PASTEURISED MILK[a]

| *Holding* | | *Coliform count* ($ml^{-1}$) | *Plate count* $ml^{-1}$ *after incubating plates at:* | |
|---|---|---|---|---|
| *Temperature* (°C) | *Time* (*days*) | | *32°C, 2 days* | *5°C, 10 days* |
| | | *Sample A* | | |
| 2 | 2 | <1 | 1 800 | 170 |
| 2 | 6 | 3 200 | 360 000 | 420 000 |
| 5 | 6 | 1 800 | 1 200 000 | 1 200 000 |
| 10 | 3 | 52 000 | 1 300 000 | 1 300 000 |
| 15 | 2 | >3 000 | 500 000 000 | 230 000 000 |
| | | *Sample B* | | |
| 2 | 2 | 1 | 1 700 | <100 |
| 2 | 18 | <1 | 10 000 | 23 000 |
| 2 | 32 | <1 | 5 500 000 | 11 000 000 |
| 5 | 16 | <1 | 1 700 000 | 1 600 000 |
| 10 | 9 | <1 | 85 000 000 | 16 000 000 |
| 15 | 3 | 17 000 000 | 47 000 000 | 630 000 |
| | | *Sample C* | | |
| 2 | 13 | <1 | 13 000 | <300 |
| 5 | 16 | <1 | 290 000 | <10 000 |
| 10 | 7 | <1 | 1 800 000 | <10 000 |
| 15 | 4 | <1 | 18 000 000 | <1 000 000 |

[a] Adapted from data of Nelson and Baker (1953).

*Detection of Psychrotrophic Bacteria in Milk*

The customary procedure for the enumeration of psychrotrophic bacteria is to incubate agar plates at 7°C for 10 days but, if the contamination level is high, the product may be spoiled or very close to it before the laboratory results are available. Examination of the plates at intervals during incubation may detect some colonies well before 10 days and, thus, permit earlier evaluation of the situation. The absence of colony forming units from the 1 or 2 ml quantities customarily plated has not correlated well with keeping quality, and plating for psychrotrophic bacteria in freshly pasteurised milk is generally conceded to be of little value in controlling defects.

The standard plate count with incubation at 32°C for 48 ± 3 h is of little value as an indicator of psychrotrophic bacteria in freshly pasteurised milk, for while a large portion of the organisms of this group will grow on plates incubated at this temperature, some will not; 25 or 28 °C being more suitable. Thus, at 32 °C, nearly every colony will be produced by a mesophilic, thermoduric organism of little or no importance to the keeping quality of the product, and if colonies are produced by psychrotrophic bacteria, they will be lost among the colonies of other micro-organisms since they have no specific distinguishing visible characteristics. Some workers have suggested flooding the plates with α-naphthol or some other reagent to detect the oxidase-positive colonies of the *Pseudomonas* spp., and similar types, since the thermoduric organisms making up most of the population in freshly pasteurised milk are oxidase-negative. Media containing crystal violet, neotetrazolium chloride, penicillin, alkylaryl-sulphonate and various combinations of these have been used to inhibit, essentially, all but the Gram-negative bacteria. Combinations of these selective agents, with incubation in the 21–32 °C range, have been advocated as a means of estimating the possible psychrotrophic bacteria in a much shorter time than usual, but none of these modifications has been adopted as a standard procedure.

**Coliform Bacteria**

As applied to dairy products, this group of bacteria comprises the aerobic and facultatively anaerobic, Gram-negative, non-spore-forming rods which ferment lactose with the production of acid and gas at 32 °C within 48 h. The genera *Escherichia*, *Enterobacter* (formerly *Aerobacter*) and *Klebsiella* are the typical organisms of this group, although a few lactose-fermenting species of other genera may also be included.

Detection of this group in water supplies has been used widely as an index of possible contamination with micro-organisms of faecal origin, but this is not the situation with dairy products. In dairy products, the coliform tests are not intended to detect faecal pollution specifically, or to identify *Escherichia coli*. Experience has shown that coliform bacteria are not usually found in properly pasteurised milk immediately following heat treatment. Therefore, tests for coliform bacteria are used to detect significant post-pasteurisation contamination. In most cases, the tests are made on the product from the final container but, in some instances, samples taken at various stages along the processing line are tested to assist in pin-pointing areas of contamination. Heat-resistant coliform bacteria have been isolated, but experience indicates they are relatively unimportant in pasteurised milk. Excessive numbers of coliform bacteria in raw milk might, under unusual circumstances, permit a few cells to survive pasteurisation. Such a situation would be quite unlikely with a properly controlled milk supply. The critical literature review by Buchbinder and Alff (1947) can be consulted for more information concerning heat-resistant coliform bacteria.

Coliform tests have several advantages for milk control. Results are available in 24 h when solid media are used, and preliminary results are available at this time when liquid media are employed. In addition, the combination of inhibitory and selective properties of the media means that as few as one coliform organism amongst great numbers of other bacteria can usually be detected. One exception to this rule is when the population of other Gram-negative organisms is large, and they grow so extensively on the same medium that they prevent the production of typical colonies by the coliform bacteria. Since coliform bacteria are widely distributed, and are quite common in the environment in which milk and milk products are handled, even a low level of contamination is likely to include some bacteria of this group.

The common standard is that coliform bacteria shall not exceed 10 $ml^{-1}$ in pasteurised milk. This is a very lenient standard and one which a well-run plant should be able to better by a wide margin. Probably one reason for an enforcement standard at this level is the relative inaccuracy of population estimates at lower levels. When solid media are used, one or two plates on which five or less coliform colonies develop gives a result with considerable probable error. When limiting dilution procedures are employed, the probable error is even greater, even when five or more tubes are used at several dilution levels. However, in routine plant control, the larger number of tests run each day, and the more frequent sampling, tends to permit a reasonably accurate estimate of contamination with coliform bacteria.

The absence of coliform bacteria in 1 ml samples of product would be a desirable objective for a well-run plant, and splitting a 10 ml sample between three plates permits a more stringent standard to be employed. Positive results on a 1 ml sample should call for correction of an undesirable situation, but it must be emphasised that the absence of coliform bacteria in 1 ml quantities of product will not assure long-term keeping quality. Sample *A* in Table II illustrates this point.

Coliform bacteria allowed to grow to numbers of a million or more per ml may cause defects. These defects may include ropiness, grassiness, unclean and medicinal odours, and bitterness. Extensive growth in raw milk may make it unsuitable for processing. In pasteurised milk, keeping quality may be affected, particularly if the milk is held above 7·2 °C (45 °F). Ropiness in refrigerated pasteurised milk may be due to post-pasteurisation contamination with strains of *Enterobacter aerogenes*. The presence of enteropathogenic *E. coli* has caused human illness from other dairy products, but pasteurised milk has not been implicated.

Poor cleaning and inadequate microbiocidal treatment of equipment with which the pasteurised milk comes in contact undoubtedly are the most important sources of contamination with coliform bacteria. Poor design and maintenance contribute to this problem. Unsanitary practices by personnel, such as assembling equipment without proper washing and microbiocidal treatment of the hands, holding gaskets and miscellaneous small parts in the mouth or in pockets, and carelessness in placing equipment either on the floor or in other areas where it may be splashed upon, are additional sources of contamination.

A problem with coliform bacteria in pasteurised products is almost always traceable to something in the processing plant that needs to be corrected, and to blame the milk producer is to procrastinate on the solution to the situation.

## MODIFIED MILK PRODUCTS

Included in this category are chocolate, reconstituted, reduced-fat, 'filled', lactase-treated and 'sweet acidophilus' milk products which have much the same microbiological characteristics as does fluid market milk. Egg-nog could also be included. Some are labelled 'drinks' because their composition does not qualify them to be labelled 'milk', and other names have been employed to avoid confusion with similar, but unmodified, products. All of these products spoil in essentially the same way as fluid milk when post-pasteurisation contamination with psychrotrophic bacteria occurs but, because these products are frequently processed in smaller lots for which the equipment may not be handled as satisfactorily as for larger scale operations, spoilage by psychrotrophic bacteria is more of a problem than it is with fluid milk. Some tendency for the period between production and consumption to be longer also contributes to problems with keeping quality.

### Chocolate Milk and Chocolate-Flavoured Drinks

These products customarily have approximately 5% added sugar, a cocoa-derived flavouring material (frequently with some non-chocolate fortification), and a stabiliser to minimise settling out of particulate materials. The sugar ordinarily adds no significant organisms, although it is not sterile. Some sugar is purchased with specifications of maximum permissible microbial content, particularly to ensure low levels of spore-forming bacteria. The chocolate flavouring may be liquid or powder, and usually poses no microbiological problem; some purveyors provide cocoa-derived products with a guaranteed low microbial content. A slightly slower growth of some organisms found in milk has been reported when cocoa-containing material has been added, but this certainly does not prevent growth and defect production in chocolate milk. The added solids increase the temperature needed for equivalent microbial destruction, so the usual requirement for pasteurisation is to increase the temperature employed by 3 °C (5 °F) above that for fluid milk. Post-pasteurisation contamination probably led to the outbreak of *Yersinia enterocolitica* gastro-enteritis epidemiologically attributed to chocolate milk, although this was not proven.

**Reconstituted Milk**

In numerous areas of the world where climatic and/or economic conditions are not conducive to the local production of milk, dried or concentrated components are brought in. Both the saving in transportation costs and the greater keeping quality of the dried products dictate the use of these materials. Following combination in the desired ratios, including the necessary water, processing proceeds much as for normal fluid milk, and the microbiological problems are essentially identical. Particular attention must be paid to adequate protection from post-pasteurisation contamination.

**Filled Milks**

These are products in which the milk fat has been replaced by other fats of similar physical characteristics. The lower cost of the substitute fat, and the alleged advantages of a vegetable fat lower in cholesterol and higher in unsaturated fatty acids, have been factors in the introduction of these products in some areas where they are legally permissible. Studies at the Arizona Agricultural Experiment Station have shown that the commercial products have the same microbiological problems as do the unaltered milk products while, in addition, the lower volume produced and the longer average time on the shelf at the retail outlet, contribute to more problems associated with microbiological keeping quality.

**Lactase-Treated Milk**

Intolerance to lactose is apparently found in a significant fraction of the human population, particularly among adults, with appreciable differences between ethnic groups. The gastro-intestinal complaints include gas, bloating and diarrhoea, and a reduced production of the lactose-hydrolysing enzyme, lactase or β-galactosidase, by the person is involved. This enzyme is also produced by a variety of micro-organisms, including the lactic streptococci and lactobacilli, *E. coli*, *B. subtilis*, *Kluyveromyces lactis* and various species of *Aspergillus*. The isolated enzyme has been added, as such, to milk or milk products, or it has been immobilised on a variety of carriers and the fluid milk passed over the immobilised enzyme. A finite time for enzyme activity is required, depending on such factors as temperature, pH, characteristics of the particular enzyme and characteristics of the product treated. Conditions which will minimise microbial development during treatment are essential, both to prevent excessive growth in the product, and to avoid problems with the immobilised enzyme. Treatment followed by pasteurisation is probably the sequence of choice, although enzyme preparations are available for adding to the pasteurised product, which is then held under refrigeration to minimise microbial growth during the period of some hours required for significant enzyme activity. The topic is too broad for detailed treatment here.

**'Sweet Acidophilus' Milk**

This product results from the introduction into normal pasteurised milk, prior to final packaging, of a suspension of *Lactobacillus acidophilus* grown in other liquid media, centrifuged off, appropriately rinsed and resuspended. The taste and pH of the milk are not altered, so removing one of the major criticisms of fermented acidophilus milk. In addition to being intestinally implantable, the strain used must remain viable in the refrigerated milk for a reasonable period of time. The usual criterion is that the viable population should exceed $2 \times 10^6\,ml^{-1}$ at the time of consumption. Appreciable differences in survival of *L. acidophilus* were noted among the three brands of 'sweet acidophilus' milk tested at the Arizona Agricultural Experiment Station (Young and Nelson, 1978). Psychrotrophic bacteria eventually developed in all samples, as would be expected from pasteurised milk which had been slightly contaminated after pasteurisation, but no evidence was obtained that development of the psychrotrophic bacteria influenced the viable count of *L. acidophilus*. An enriched medium and incubation of plates in a nitrogen–carbon dioxide atmosphere were necessary for accurate enumeration of the several strains of *L. acidophilus*.

**Egg-Nog**

This product contains added egg, sugar, flavouring and frequently added milk solids. The microbiological quality of the ingredients must be controlled adequately, and because of the high solids content, and probably to some degree because of the possibility that the eggs could contain *Salmonella*, the pasteurised milk ordinance states that pasteurisation shall be at 69 °C (155 °F) for 30 min, 80 °C (175 °F) for 25 s, or 83 °C (180 °F) for 15 s.

## BOILED MILK

In many situations where milk processing technology has not been highly developed, boiling of raw milk as soon as possible after delivery to the consumer is practiced. By raising the temperature to essentially 100 °C (212 °F) (the boiling point being influenced by altitude and solids concentration), all micro-organisms except some of the spore-forming bacteria in the spore state will be killed, including those which might directly cause human disease. Care must be taken to avoid growth of enterotoxigenic staphylococci prior to heat treatment, because staphylococcal enterotoxin is not inactivated by boiling. The boiling should be done with minimum delay or, alternatively, the unprocessed milk must be held below 10 °C (50 °F).

Boiled milk will have a limited microbiological keeping quality, particularly if held without adequate refrigeration, because the bacterial spores which survive the heat treatment will germinate at typical room temperatures. Many of these organisms are markedly proteolytic and, thus, can cause pronounced off-flavours and may modify the physical characteristics of the milk. Some produce gassiness, while others may cause coagulation by acids produced and/or coagulating enzymes.

Contamination subsequent to heating is a practical problem with boiled milk. Portions of the product may be incompletely heated, as with product which has splattered well up on the sides of the container, but probably more important is the common use of containers and equipment, such as dippers and cups, which have not undergone adequate microbiocidal treatment. Under most circumstances, particularly where refrigeration is poor or lacking, boiled milk should be consumed in less than 16 h.

## THE SPREAD OF DISEASE THROUGH MILK

In recent years, very few milk-borne outbreaks of human illness have been reported in the United States, and the last summary of food-borne illness available, for the year 1977 (USDHEW, 1979), listed no outbreak attributed to milk, either raw or pasteurised. In considerable contrast to this, during the period 1938–1950, 401 outbreaks involving 16 232 persons were reported due to milk and milk products. Despite such notable progress, considerable attention must continue to be given to the possibility that milk may serve as a vehicle for disease transmission. In the present pattern of marketing, product from a single processing plant will reach large numbers of people frequently scattered over a considerable geographical area. Thus, many people could be affected by a contaminated or improperly handled lot of milk. In some areas where disease control and technology are less developed than in northern Europe and the United States, milk-borne disease remains a major problem.

Some milk-borne diseases are caused primarily by micro-organisms that infect the cow, or other milk-producing animals, and gain entrance directly from the animal source, either being present in the milk as it is drawn from the udder, or coming from the immediate environment of the animal. Tuberculosis, brucellosis, some forms of salmonellosis, and septic sore throat or scarlet fever are examples of this type of illness. The rickettsia of Q fever may be present. *Staphylococcus aureus* is frequently present, being a common cause of mastitis, and enterotoxigenic strains of staphylococci may grow enough to produce a level of toxin able to cause severe gastroenteritis among people who consume such milk. In areas where foot and mouth disease is endemic, attention must be directed to this disease. Anthrax and Johne's disease are of little importance from the standpoint of human illness spread by milk, but they must receive some consideration. Great advances have been made in the United States and a number of other countries in reducing, to very low levels, the incidence of tuberculosis and brucellosis in dairy herds. Much remains to be done in reducing levels of mastitis due to staphylococci, *Str. pyogenes* and other micro-organisms which may be involved in human illness.

A second group of milk-borne illnesses is caused by pathogens for which humans are the principal reservoir; typhoid, diphtheria and poliomyelitis are examples. These gain entrance to milk directly from humans, from utensils, or from water contaminated from human sources. Milk can probably serve as a mechanical carrier of almost any bacterium or virus causing illness in man. However, periodic testing of milk handlers to detect those who might be shedding one or more pathogenic micro-organism has not proved a fruitful means of combating milk-borne illness.

The marked reduction in milk-borne disease over the years is attributable to: (1) the widespread use of pasteurisation, which is designed to kill the most resistant milk-borne organisms responsible for human illness; (2) a marked reduction in the incidence of many of these diseases among the animal and human reservoirs of the causative micro-organisms; (3) better equipment and understanding of its use, so that processing is carried out properly and chances for contamination during, and subsequent to, processing are minimised; and (4) greater development and more intelligent use of the regulatory function. Continued improvement in the application of these principles can be depended upon to decrease still further the incidence of milk-borne illness.

## REFERENCES

APHA (1978) *Standard Methods for the Examination of Dairy Products*, 14th edn, American Public Health Association, Washington, DC.

BUCHANAN, R. E. and GIBBONS, N. E. (Eds.) (1974) *Bergey's Manual of Determinative Bacteriology*, 8th edn, Williams and Wilkins Co., Baltimore, Md.

BUCHBINDER, P. and ALFF, E. C. (1947) *J. Milk Fd Technol.*, **10**, 137–48.

GORDON, W. A., MORRIS, H. A. and PACKARD, V. (1980) *J. Fd Protection*, **43**, 58–64.

LAW, B. A. (1979) *J. Dairy Res.*, **46**, 573–88.

LAWTON, W. C. and NELSON, F. E. (1954) *J. Dairy Sci.*, **37**, 1164–72.

MEHTA, R. S. (1980) *J. Fd Protection*, **43**, 212–25.

NELSON, F. E. and BAKER, M. P. (1953) *J. Milk Fd Technol.*, **17**, 95–100.

PRICKETT, P. S. (1928) Thermophilic and thermoduric micro-organisms with special reference to species isolated from milk. V: Description of spore-forming types, *NY Agr. Expt. Sta.* (*Geneva*) *Bull.*, 147.

SLATTER, W. L. and HALVORSON, H. O. (1947) *J. Dairy Sci.*, **30**, 231–43.

STOLOFF, L. (1980) *J. Food Protection*, **43**, 226–30.

THOMAS, W. R., REINBOLD, G. W. and NELSON, F. E. (1963) *J. Milk Fd Technol.*, **26**, 357–63.

THOMAS, W. R., REINBOLD, G. W. and NELSON, F. E. (1966*a*) *J. Milk Fd Technol.*, **29**, 156–60.

THOMAS, W. R., REINBOLD, G. W. and NELSON, F. E. (1966*b*) *J. Milk Fd Technol.*, **29**, 182–6.

USHDEW (1979) *Foodborne and waterborne disease outbreaks, annual summary, 1977*, Center for Disease Control, United States Department of Health, Education and Welfare, Atlanta, Georgia.

USFDA (1978*a*) *Grade A pasteurised milk ordinance*, Superintendent of Documents, US Government Printing Office, United States Food and Drug Administration, Washington, DC.

USFDA (1978*b*) *Recommended sanitation ordinance for condensed and dry milk products and condensed and dry whey used in Grade A pasteurised milk products*,

Superintendent of Documents, US Government Printing Office, United States Food and Drug Administration, Washington, DC.

USFDA (1978*c*) *Fabrication of single-service containers and closures for milk and milk products*, Superintendent of Documents, US Government Printing Office, United States Food and Drug Administration, Washington, DC.

YOUNG, C. K. and NELSON, F. E. (1978) *J. Fd Protection*, **41**, 248–50.

# 6

# The Microbiology of Dried Milk Powders

H. R. Lovell
*School of Food Studies,*
*Queensland Agricultural College, Lawes, Australia*

It might be thought that dried milk products are of relatively recent origin, however, this is not the case. The earliest reference is by Marco Polo who in the 13th century visited the court of Kublai Khan. He refers to the drying of milk into a paste by simmering the milk and continuously removing the fat-rich component which rose to the surface. This portion was used to make butter and kept separate. The skim-milk remaining was concentrated further with the final stage achieved by solar drying. On expeditions each man took some ten pounds of the dried milk with him. He would reconstitute the product by placing some of the dried material in his leather water bottle and adding water. Mixing of the product took place as he rode along until finally he had produced a semi-solid product which would constitute a major meal. In 1839 the French traveller Rubruquis confirmed Marco Polo's observation, when he reported that the Mongols produced a similar product which they called 'Kurut'. The microbiological quality of these products would have been interesting, particularly when reconstituted.

In 1810 Nicholas Appert concentrated milk slowly until it formed a paste, and then completed the drying process by agitating the product with hot air, thus producing the forerunner of the fluid bed. Grimwade, an Englishman, in 1855 patented a process for producing desiccated milk. Sodium carbonate was added to whole milk which was then condensed in an open pan to form a paste. The process then continued with the addition of sugar, and finally the mixture was passed through two rollers to give a powdery end-product.

Developments after this date included Samuel R. Percy's patent for the spray process, and the roller drying equipment of Just and Hatmaker.

Diversified milk products were also to the fore with the malted milk patented by William Horlick in 1883; a commercial reality four years later. Thus, we have some 80 years of experience of the drying of milk to which to refer. The record to date, viewed as a whole, is to the credit of the industry, with comparatively few recorded incidents of a microbiological nature relating to dried milk. It could be argued that the nature of the product, i.e. the low moisture levels, helped to ensure product safety. The early use of dried milks as baby foods undoubtedly played a major role in production and quality control. Few manufacturers could survive any outbreak of infection amongst babies and infants which involved their product. Equally, only manufacturers with previous experience in processing ventured into drying operations due to the scale and the initial cost. The dairy industry has always been conscious of the microbiological hazards associated with milk and, thus, it is hardly surprising that dried milk has had a clean bill of health in the historical sense.

In recent years, the development of new microbiological techniques, the significance of salmonellae and staphylococcal food poisoning, the increasing range of micro-organisms found to cause infections, and the newer processes, such as 'instantising', have all helped to focus attention on the microbiology. The biggest influence, however, might well be argued to be the growth in specifications. The upsurge and evolution of trade between countries, as well as the pursuit of consistent standards of quality by food manufacturing companies for their products, has focused attention on this aspect. The influence of governmental agencies, standards associations and international federations have aided and encouraged this development. The last two decades, therefore, have seen an escalation of activity in this area and, at the same time, the deficiencies in our knowledge relating to the basic techniques of microbiological analysis for milk powder have been highlighted.

## MANUFACTURING PROCEDURES

At present, two basic processes are employed in the drying of milk products, roller drying and spray drying. Other procedures are used for specialised milk products, e.g. compounded beverages, and the introduction of 'instant' products has seen the incorporation of a number of post-processing operations. These post-processing procedures have increased the microbiological hazards but, nevertheless, have not given rise to any insoluble commercial problems.

**Roller Drying**

The raw milk is received, bulked and standardised; some manufacturers pasteurise the milk after bulking, but prior to standardisation. After standardisation, the milk is fed to the evaporators for pre-concentration prior to drying. Preconcentration is not essential but is widely practised as it is a more economic method for the removal of moisture.

The level of total solids in the preconcentration stage varies from manufacturer to manufacturer. It may be as low as 16–18 % total solids (Hall and Hedrick, 1971) or up to 26–28 %. The concentrated milk is then fed to the roller dryers. A range of roller dryers are employed in the dairy industry, and they vary from single to double drum types (Fig. 1) operating either at atmospheric pressure or under vacuum conditions. The method of supplying the concentrated milk ranges from a central sparge pipe centred above the drums, to twin rollers transferring the concentrate from a trough running the whole length of the machine.

The drums are supplied with steam up to 150 °C, and rotate at speeds between 12 and 17 rpm. The milk dries rapidly on the hot surface of the drum, and is removed by a sharp blade ('doctor' knife) in close contact with the surface. The powder comes away from the drum in a crinkly film

FIG. 1. A typical battery of roller-drying plants used for the production of high quality milk powders. Reproduced by courtesy of St. Ivel Ltd, Wincanton.

resembling crêpe paper, and falls into the collecting troughs situated parallel to the drums (see Fig. 2). The collection troughs may contain a coarse pitch screw, of a diameter comparable to the width of the trough, which moves the coarse material to one end. The product then exits into a conveyor system running transversely to the dryer. Alternatively, the coarse material may be removed by hand, using large metal scoops and transferred to a portable bulk bin.

FIG. 2. An individual machine; the milk is pumped into the central well from an over-head pipe (right), and the 'sheets' of dry product can be seen leaving the hot rollers. Reproduced by courtesy of St. Ivel Ltd, Wincanton.

Whichever procedure is employed, the product is usually passed through either a Kek mill or brush sifter. This breaks the product into a relatively uniform powder prior to packing. The powder may then be packed into 25 kg sacks or, alternatively, into retail units. Prior to the latter packing procedure, however, it is frequently the practice to hold product in an intermediate bulk container or silo. Where powder is packed into sacks, it may be passed through a bag 'squasher' to compact the bag; this has the effect of removing some of the occluded air and permits easier stacking.

The packaging of the retail units (Lovell, 1972) can be a highly sophisticated, totally automatic operation, or it can involve a considerable amount of manual input. A great deal of roller-dried milk powder is used for baby foods, and the retail packs frequently require the insertion of a measuring spoon in addition to reconstitution instructions. The packs are finally coded, collated and shrink-wrapped into larger market units.

**Spray Drying**

The raw milk is received, bulked and standardised for fat content, but the practice of pasteurising milk varies between manufacturers, and is by no means universal. However, there is a preheating stage prior to evaporation, and this will vary according to the type of powder being manufactured. In respect to non-fat dry milk, two principal types of powder are produced, low-heat non-fat dry milk and high-heat non-fat dry milk. Low-heat powder usually receives a pasteurisation treatment of 72 °C for 15 s, whereas high-heat powder in addition to pasteurisation, receives a further heat treatment of 85 °C for 20 min.

In the case of whole milk and non-fat milk not designated as above, the general practice is to heat the milk to 88 °C or above prior to concentration. The concentration of milk, be it whole or non-fat, is usually carried out in multi-effect evaporators, with or without a finishing evaporator to bring the total solids up to the desired level. The developments in evaporator design over the last 10 to 15 years have resulted in some highly sophisticated equipment, designed principally to achieve the maximum savings in heat utilisation, and the incorporation of UHT steam injection heads is not unusual, particularly where a high grade final product is required.

Following concentration, the product is ready for drying. In some cases a balance tank may be inserted between the concentration and drying stages, or the dryer may be fed direct. Essentially two systems of spray drying are employed at present. The jet or nozzle type spray dryers are favoured in the USA and by US-based companies, whilst those companies employing the rotary atomiser are more commonly found in Europe. There have been attempts to make the two systems available in the one unit but, thus far, this has not been completely successful, commercially.

Milk to be dried by the jet-spray drying process is first preheated, and then fed to a high pressure pump, similar in design to the familiar homogeniser but without the homogenising valves. The concentrated milk is then pumped under pressure to a series of jets within the drying chamber. The milk is sprayed in a thin conical sheet from each jet, and the accompanying hot air flow breaks the thin film causing droplet formation, and a fine powder results. The heavier powder falls to the base of the chamber, which is usually rectangular in shape, and is removed continuously by scrapers, whilst the lighter powder ('fines') is carried out in the now moisture-laden hot air stream. The 'fines' are removed from the hot air stream either by a cyclone, or series of cyclones, in which the fine powder is separated by centrifugal action; or it is removed by cloth filters of the tubular type. The cloth filters are usually agitated at intervals to remove the build-up of powder.

In the case of spray-drying with the rotary atomiser, the concentrated milk is fed direct, or via a balance tank, to the atomiser. The atomiser, in general terms, comprises a high-speed electric motor driving a shaft on which is fixed a circular atomiser disc; the design of this disc varies with each spray dryer manufacturer. Essentially the disc has a number of slots or

FIG. 3. A spray dryer equipped with a pneumatic transport system, suitable for production of dried milk products such as ordinary skim-milk, whey powder and baby food. Powder produced on this type of plant tends to be dusty, as there are no agglomerates, but powder bulk density is high. Reproduced by courtesy of Niro Atomizer Ltd, Denmark.

holes around its periphery through which the concentrated milk exits at high speed whilst the disc is spinning. The speed of the disc and the manner in which the concentrate leaves the disc leads to the break-up of the milk into fine droplets. The atomiser is mounted in a large chamber to which hot air is supplied, and the droplets lose moisture rapidly to become a powder. The powder is removed with the hot air stream from the main chamber and is passed through wide diameter trunking to the primary cyclones (see Fig. 3). Here the heavier powder separates, whilst the lighter powder ('fines') is carried onward with the hot air. The secondary powder removal system may comprise cyclones or, alternatively, bag filters. The powder is usually supplied to a small silo for bagging off, or to a large bulk silo pending packing or further processing.

The introduction of 'instantising' has brought further additions to the

FIG. 4. A spray dryer equipped with vibro-fluidisers (vibrated fluid beds) in series, and used to produce instant skim-milk powders directly. It is also suited for production of non-caking whey powder of high fat content, protein powder, agglomerated whole milk powder, instant beverage whiteners and baby food formulas. This layout involving fluid beds in series enables the spray dryer to be operated as a two-stage process. Two-stage drying increases product capacity per spray drying chamber, improves powder quality, and decreases overall heat consumption since the spray dryer is operated at lower outlet temperatures. Reproduced by courtesy of Niro Atomizer Ltd, Denmark.

spray drying system, and has added to the microbiological hazards. 'Instantising' is the process of forming agglomerates of powder which improves the wetting and dispersing properties of the basic product. The process may involve the recycling of a newly made powder back to the atomiser head so that the powder is brought into intimate contact with the moist droplets, thus forming agglomerates; this is the so-called straight-through process. Alternatively, the powder may be introduced into a turbulent, warm, moist atmosphere where collision between semi-moist particles leads to the formation of agglomerates. In either process, the resultant granules require a further and final drying stage which is usually carried out by one or two fluid-beds (see Fig. 4). The introduction of lecithination for the production of instant whole milk has added a further complication to the spray drying process, and doubtless other developments will follow.

The packaging of spray-dried milk powder is similar to that of roller-dried powder (Lovell, 1972). Bulk packaging may be in 25 kg sacks (Fig. 5) or even larger re-usable flexible containers, or into retail unit packs.

FIG. 5. It is essential that milk powders are stored under cool, dry conditions, and the store illustrated is an excellent example of a system that makes maximum use of a storage facility without compromising an 'ease of access' or ventilation. Reproduced by courtesy of Unigate Foods Ltd, St. Erth.

'Instantised' products are also packed in sacks or fibreboard drums, and in unit packs including cartons, tins, convoluted, spiral-wound cardboard drums, or small multi-wall paper bags. Retail units are collated into packs of 20 or more containers, shrink-wrapped and palletised.

### Alternative Drying Procedures

A number of milk products have used established drying procedures over many years, and many have not been modernised for fear of inducing a noticeable change in the product. Vacuum ovens are widely used for malted milk products, for example, and rotary tunnel dryers have long been used for casein. Microbiological data relating to these processes, and the end products, remains with the companies concerned, and little is known of the efficacy of these processes in microbiological terms.

New processes are introduced from time to time. The BIRS process (Waite, 1964) in the 1960s, which used a gentler drying process proved uneconomical for milk. The recent foam mat process results in a porous product and, consequently, the end-product shows improved 'wettability' characteristics.

## MICROBIOLOGICAL ASPECTS OF PROCESSING

The microbiological aspects of processing, from a routine point of view, are of direct concern to factory managers, quality controllers and, particularly, to those directly supervising the process but, all too frequently, microbiological factors only gain prominence, resulting in the implementation of control procedures, by default rather than as the outcome of careful planning.

The classic work on the microbiology of the spray drying of milk is the Staplemead experiment. This was a series of trials carried out on a commercial evaporation and spray drying plant at Aplin and Barrett's factory at Staplemead, England. The work was carried out by Mattick, Crossley and colleagues in 1942, and was directed towards extending the storage life of whole milk powder; nevertheless, it encompassed a number of important microbiological aspects. (Mattick, *et al.*, 1945; Crossley and Johnson, 1942).

The classic paper on the microbiology of roller-dried milk powders is that of Higginbottom (1944). Whilst these references may be regarded as vintage publications, the basic thinking and conclusions have withstood the test of time.

In examining the influence of the various stages in microbiological terms, it is necessary to examine each process stage by stage.

**Roller Drying—Microbiological Considerations**

The roller drying process quickly earned itself a reputation for being the safest process in microbiological terms, and within the United Kingdom it was for many years the only process permitted for the production of National Dried Milk. Its microbiological efficacy lay in the high temperature to which the product was exposed when in contact with the hot rollers in the drying stage. Whilst its popularity is diminishing within the dairy industry in favour of spray drying, it is prudent to record the microbiological aspects.

The prime consideration in examining raw milk is the determination of the numbers of thermoduric micro-organisms. This is the only meaningful numerical determination which can be related to the finished product. Naturally, the total colony count and the determination of coliforms are of significance in assessing the general hygienic quality, but they have less influence on the quality of the final product.

The rapid cooling of the milk prior to further treatment is important in allaying any microbial increase, but it is commonplace to pasteurise milk as quickly as possible prior to further processing. Clarification is not thought to contribute to any serious reduction in microbial numbers, but is effective in removing larger cellular matter and macroscopic material.

Homogenisation is usually incorporated in the roller drying process in order to reduce the occurrence of free fat on reconstitution. This will also have an effect on the total colony count due to the breaking-up of bacterial clumps (Trout, 1950). This can be confusing in the case of a homogeniser in which there is seepage of contaminated cooling water past the pistons due to worn or defective chevron rings. This can be observed by any significant change in microflora from milk taken previously in the run.

The standardising stage is unlikely to create any difficulties, and the next stage of significance, microbiologically is evaporation. The necessity to take a vigilant approach towards the hygiene of evaporators and their operation cannot be overstressed. Insofar as thermoduric micro-organisms are concerned, evaporators can act as incubators, and even continuous fermenters, with massive proliferations of micro-organisms taking place. Detailed reference to the salient microbiological features of evaporators will be found under spray drying process considerations.

Following the evaporator, it is necessary, in the case of roller drying, to introduce a balance tank in order to provide a uniform flow. This can prove to be a source of contamination and covers need to be kept in place; close attention must be paid to cleaning. The balance tank then feeds direct to the roller dryers, and the commonly accepted practice is to use a perforated pipe to feed the concentrated milk into the well between the rollers. As this is an area of high temperature, there is no question of microbial growth but rather of destruction. However, the interior of the hood used for vapour removal from the dryer can provide a harbour for micro-organisms. It is an area which is warm, but not hot in the lethal sense, it has a high level of humidity, and frequently contains pockets in which microbial proliferation can take place. Furthermore, the ducting associated with the hoods is usually vented directly to the external atmosphere, and this can allow ingress of rain, insects, etc. The drying drums themselves result in a reduction of viable micro-organisms in the product, with the net result that powder taken directly from roller surfaces shows only thermoduric bacteria with aerobic spore formers predominating.

The recovery of the dried product from the rollers presents the first post-drying hazard. It is safe to assume that, from this point on, the microbiological problems are the most serious and, frequently, the most difficult to detect. Casual contamination particularly by manual contact is most likely. The collection of powder from the drying drums by mechanical means is to be encouraged, but manual methods are still employed. The milling and/or brush sifting of the product does not present any microbiological hazards, neither does bulk packaging. However, unit packaging, particularly when this involves any manual input, and the insertion of measuring spoons or leaflets by hand can result in contamination. The use of disposable plastic gloves is one method of reducing contamination where manual procedures are employed.

The routine process control of roller-dried milk powder should seek to identify hazardous areas, and samples should be taken on a regular basis. Key points for sampling are identified as follows:

(1) raw milk,
(2) milk after standardising and pasteurising,
(3) concentrated milk-balance tank,
(4) powder—ex-rollers,
(5) powder—ex-bulk silo, and
(6) powder—final package.

The tests which are normally applied to these samples are:

(1) standard colony count,
(2) thermoduric count—raw milk only, and
(3) coliform examination.

There has been little published about the routine control procedures applicable to roller-dried milk powder, and the question of the determination of staphylococci has not received the prominence which it has in spray-dried powders. Nevertheless, it is included as a routine test in many roller drying plants, and staphylococci have been reported in roller-dried powder by Hobbs (1955). The question of screening for salmonellae has also been raised from time to time, particularly as roller-dried milk powder may be used for the feeding of babies. It is generally accepted, however, that screening for salmonellae requires a high level of expertise which can rarely be justified in routine control laboratories. This argument is supported by the clean record which roller-dried milk powder has earned over the years.

The set of six samples taken on a twice weekly basis will quickly build up a bank of data on the microbiological efficacy of the particular plant. It is important not to assess microbiological results in a control situation on an individual basis, but rather to compare them with previous data. The information can then be used with greater confidence in identifying and pin-pointing hazardous areas. Such data is invaluable when high counts or other problems occur. Frequently trouble can be spotted in advance, for example, a build-up of deposits in heating plant, defective valves, etc., and such data is not easily obtained by the 'blitz' approach usually adopted when the end product only has been examined and found to be suspect. Reductions in the frequency of sampling can be made with experience, or increased as the situation demands. One important factor in carrying out routine control procedures is to avoid frequently changing methods. It is important to use one procedure and adhere to it for as long as possible if one is to make valid comparisons.

Questions are also raised concerning the effectiveness of bacteriological swabbing procedures in assessing the cleanliness of plant and, with the now almost universal use of CIP procedures, the accessibility of plant for swabbing has been seriously curtailed. There is little doubt, however, that swabbing has a role to play in keeping operatives and supervisors aware of the need for effective control over cleaning but, if the results are to reflect the true state of the plant, it has to be carried out without warning, and in practical terms, within the production environment; a practice requiring a very high level of co-operation and confidence within the plant.

**Spray Drying—Microbiological Considerations**

In considering the microbiological aspects of the production of spray-dried milk powders, it is important to recognise the influence of the roller drying process. In microbiological terms, it is analogous to the comparison between the conventional in-bottle sterilisation process and the UHT process coupled to aseptic packaging for, in both instances, the forerunner has been a process of high microbiological integrity. This has resulted in a cautious approach being adopted to the newer process, and with beneficial effect. The roller drying process produces a milk powder of low bacterial content, and has a long history of freedom from pathogens. Its employment in the manufacture of National Dried Milk by the British Government during the 1940s and 1950s reflected a confidence in the process which strengthened over the years.

Raw milk quality is of great significance, particularly if a low-heat powder is to be manufactured. Thus the milk should be of a high hygienic quality, and thermoduric organisms, in particular spore-formers, should be of a low order numerically. In addition, the increasing demand for powders with a low thermophilic count has also necessitated the acquisition of high quality raw milks. Furthermore, as will be seen later, thermophilic bacteria have ample opportunity to increase in numbers at various stages throughout the plant.

Following standardisation or separation, the milk is preheated. The Staplemead experiments identified the need for preheating, and this work subsequently formed the basis of a code of practice (Code of Practice, 1954), which has been used as an effective guide to the present day. Whilst the code of practice lays down minimum heat-treatment conditions, the practice in latter years has been to use higher temperatures for shorter times. The introduction of UHT heads into evaporator circuits, for example, has resulted in powders of high bacteriological quality. However, this is by no means commonplace, and many manufacturers merely observe the minimum conditions.

Advances in evaporation plant design over recent years have been rapid, and have resulted in sophisticated systems with the principal emphasis on economy of energy input. Three- and four-stage continuous evaporators are now the order of the day, with the principal design being the falling-film type. Preheater sections previously situated externally are now built into one or more of the calandria shells. During start-up, and in normal operation, some evaporator designs incorporate a feed-back circuit, initially for the rapid build up of concentration, but later for the effective control of solids; a modification that adds to the complexity of the plant.

The widespread introduction of multiple effects does mean that temperatures zones, which are conducive to microbial growth, can occur in plants. In addition, the complexity of plant designs is such that careful attention to cleaning procedures is crucial if a high quality product is to be produced. Hazardous points do occur on evaporators, and these include return loops, blank ends, feed control valves and the vapour ducts at the top of separators, but built-in CIP installations are singularly effective in ensuring a thorough cleaning and sanitisation of the surfaces. Where dual preheating systems are employed, it is essential to ensure that milk residues are cleared from both the circuits, and that full cleaning and sanitisation procedures are carried out.

The transfer of the concentrate from the evaporator to the drying stage can be direct, but many manufacturers prefer to use an intermediate balance tank; it had been the practice to use two balance tanks and switch over from one to the other during the run. These balance tanks provide a reserve of concentrate, thus ensuring continuity of operation, but they do represent a definite microbiological hazard (Hawley and Benjamin 1955; Keogh, 1965; 1966). The temperature of the concentrate on exit from evaporators is in the region of 45 °C, and in the balance tanks will drop slightly. The high total solids levels which are now used tend to encourage the formation of a sludge or gelatinous layer; and this material, being conducive to microbial growth, can 'seed' fresh product entering the tank. In order to overcome this problem, dual tanks were often provided and switched over during the drying run. Mesophilic bacteria, including pathogens, can and do grow in such environments, and due care must be observed. It is important that covers are retained on these tanks wherever this system is used. A further source of contamination is the measurement of density using a hydrometer, and contact with an operative's hands can result in transmission of micro-organisms, in particular, the staphylococci. Whilst the newer evaporation plants can be fitted with in-line density controllers, this is by no means always the case.

It is necessary to consider the spray drying operation from the standpoint of the two different systems. In the case of the 'jet' spray dryer, the concentrate passes through a high pressure pump, similar in design to the familiar homogeniser. The precautions which apply to the homogeniser apply equally to the high pressure pump, namely the necessity of ensuring that the pistons and associated glands are kept in good order. The cooling water supplied to the pistons must be potable, and microbiological checks should be carried out from time to time. As the supply system to the nozzles is under positive pressure, the integrity in microbiological terms is very good, and there are no particular hazards at the drying stage, particularly with the increase in inlet air temperatures over the years. At the time of making this observation, inlet air temperatures were 170 °C, or thereabouts.

In latter years, inlet air temperatures have risen to 200 °C and it has been stated that as temperatures increase so microbiological counts decrease, which is what would be expected (Chopin *et al.*, 1977; 1978; Galesloot and Stadhouders, 1968).

The recovery system for powder from 'jet' spray dryers can be either cyclone or fabric filter or both. Cyclone recovery is the safest procedure in microbiological terms. The use of fabric filters is not commonplace, as they require more attention than cyclones and are not easily cleaned. Nevertheless, they are very efficient and are used in some installations, particularly where there may be environmental considerations.

The intermediate storage of powder, whether in silos or tote bins, presents no microbiological problems. Unit packing, as with roller-dried milk powder, can pose problems with casual contamination due to hands coming into contact with the product.

In considering spray drying using rotary atomisation, the situation is similar. Product is withdrawn from the balance tank and pumped to the atomiser. Newer designs of pump using a stator of flexible material have removed many of the problems associated with earlier designs; again there are no inherent microbiological hazards within the drying chamber. Inlet air temperatures have increased steadily over the years from 160 °C to 250 °C. The filtration of air prior to heating has also improved with the use of highly retentive filters. Furthermore, the heating of air is usually sufficient to avoid any serious infection. The drying chamber presents no problems but, nevertheless, the walls should be kept in a clean condition and any build up of product quickly removed. Equally, the powder recovery systems do not present a microbiological hazard, with the possible exception of heat recovery units. Thus, various attempts have been made to recover the waste heat from the exiting air stream by the use of heat recuperators. The design of heat recuperators has followed that of the plate heat exchanger in that they consist of a series of plates between which flows alternately incoming air and exhaust air. Whilst there is no question about the energy savings, the microbiological hazards far outweigh any economic benefits. Thus, during periods of low ambient temperature, there is a problem of condensation in the exiting air stream. Whatever the efficiency of the powder separation system, small particles will continue to pass through, the net result being a moist, warm medium which is ideal for microbial growth. It could be argued that the two air streams do not come into contact with each other, but this is only true whilst the integrity of the system is secure.

Related to the problem of the air stream is the siting of the air intake. This should be in a clean area, free from dust and ideally, far removed from any activity involving the dried powder. Such areas should not be used for transient storage of other materials which can cause dust clouds, thus resulting in an ingress of micro-organisms into the drying system.

The whole question of aerial contamination has been raised by Hawley and Benjamin (1955) and confirmed by Crossley and Campling (1957). It is important, therefore, that the manufacturing area should be kept as dust-free as possible, this places special requirements on the design stages. The use of a built-in vacuum cleaning system is one reliable method for the removal of powder residues during the cleaning cycle, thus ensuring that there is no serious build up in the factory environment.

In recent years, the demand for milk powder with a very low microbial count has increased and, as a consequence, attention has been focused on the need for a high level of hygiene.

## MICROFLORA OF DRIED MILKS

The microflora of dried milks has been examined by a number of workers, notably Crossley and Johnson (1942) and Higginbottom (1944), and their data have been confirmed in the ensuing years (Keogh, 1966; 1971).

In the case of roller-dried milks, micrococci, aerobic spore-formers and *Sarcina* spp. predominate in the mesophilic range, whilst aerobic spore-formers make up the microbial population in the thermophilic range. The severe heating, which the milk receives during the drying operation on the rollers, reduces the bacterial content to a low level, and only heat-resistant spores survive to any degree. Post-drying contamination can, of course, give rise to a varied pattern of microflora.

Spray drying as a process, however, has not always enjoyed the same enviable record. Prior to the Staplemead experiments, and even following, spray-dried milk has been reported as being of varying quality (Crossley, 1962). Crossley and Johnson (1942) originally reported on the flora of spray-dried milk, which they established as being made up of thermoduric micrococci, thermoduric streptococci and corynebacteria, aerobic spore-formers and miscellaneous organisms. Following the code of practice recommendations which advised various alternative preheating procedures, the microflora centred on thermoduric streptococci and spore-formers. This is only to be expected with the selective action imposed by the various heat treatments, the most severe of which involved heating to 88 °C.

Latterly there has been a move towards the ultra-high temperature heat treatment of milk for processing into milk powder. Such action results in near-sterile powder with only casual post-heat treatment contaminants being present. In addition, inlet air temperatures in the spray drying operation have risen, but even temperatures above 200 °C fail to remove all spore-formers.

The routine microbiological control of spray-dried milk powder should be directed at the pattern of the microflora throughout the production run; again the procedure should aim at sets of samples being taken at regular intervals. The sampling points might well be made up of the following:

(1) raw milk,
(2) milk prior to concentrating,
(3) concentrated milk immediately prior to atomisation,
(4) powder ex-primary cyclones, and
(5) powder ex-final pack.

If 'instantising' is incorporated, then in the case of straight-through 'instantising', an additional sample at the fluid bed stage is sufficient. In the case of re-wet 'instantising', then the following sampling pattern should be considered:

(1) powder,
(2) agglomerated material at the fluid bed, and
(3) final package.

The tests carried out on the samples normally include:

(1) standard colony count,
(2) thermoduric count,
(3) presence of *Staphylococcus aureus*, and
(4) coliform examination.

Again, the detection of *Salmonella* spp. is not recommended as a routine procedure to be carried out in the control laboratory.

The use of plant swabs has some merit in a spray drying operation, particularly in those areas where powder particles can accumulate and take up moisture. The cleaning of spray dryers varies considerably both in degree and frequency, but it is not unusual for dryers to receive only two major cleaning runs in a production season of several months. Such practice is neither to be condoned nor encouraged. Frequent cleaning on the basis of regular usage is essential, as should be the regular inspection of plant for essential repairs. Many of the routine microbiological problems can still be traced to poor seams, pockets caused by poor design or bad welding, and these need to be tracked down and rectified.

**Bacterial Counts**

The quantitative assessment of the bacterial content of milk powders has proved to be an area of divergent opinions. In seeking to lay down advisory standards for the bacterial content of milk powders, all the vagaries of trying to obtain a reproducible microbiological technique have come to the fore (Wallgren, 1964).

Roller-dried milk powder has previously had a much lower bacterial content than spray-dried milk powder and, for this reason, was highly favoured as an infant food during the years of the Second World War and its aftermath. The standards laid down for National Dried Milk contracts in the United Kingdom required a maximum of not more than 5000 bacteria per gram after three days incubation at 37 °C, and not more than 1500 bacteria per gram after two days at 55 °C. The vast majority of manufacturers had little difficulty in achieving these standards, and indeed it was a rare event when the thermophilic count exceeded 1000 bacteria per gram.

The picture in relation to spray drying is by no means as clear-cut, and bacterial counts for spray-dried milk powders are quoted between thousands and millions per gram. Hence, it has become accepted that spray-dried milk powders will have higher bacterial contents than roller-dried milk powders. This attitude is somewhat unfortunate, as responsible manufacturers with a conscientious and vigilant approach to the whole of the spray drying operation can achieve very low bacterial contents indeed. This is a very necessary objective, especially in those cases where the powder is destined for incorporation in some other product, for example ice cream, canned products or compounded baby foods. Various standards have been suggested, but 50 000 organisms per gram appears to have been almost universally accepted (Murray, 1975; Cox, 1970). However, the situation is by no means clear, and there is much controversy over the method by which the organisms are determined. Thus the method of the American Dry Milk Institute (ADMI) proposes 2 days at 35 °C, whereas the International Dairy Federation method (IDF, 1970) cites 3 days at 30 °C, and not only has the temperature of incubation been considered, but also the temperature of reconstitution. However, Bockelmann (1969) suggested that, although the IDF method required reconstitution at 50 °C and the ADMI method proposed 45 °C, the temperature of incubation had the major influence.

Various workers have been preoccupied with the liquids used for reconstitution and dilution. Lithium hydroxide was an early recommendation because it aided dissolution of the powder, but it was then shown to be bacteriocidal (Garrison, 1946). The next move was to sodium citrate and then to water, all of which were claimed to give higher and, therefore, supposedly truer counts. Such developments make it increasingly difficult to assess historically the true picture, in microbiological terms, of the spray drying process. At least Bockelmann (1969) and Lubenau and Mair (1978) must be credited with a careful comparison between the newer, and more widely accepted, ADMI and IDF procedures.

The important fact to bear in mind is that each plant has its own bacteriological flora, and whilst microbiological counts tell little when viewed in isolation, the emergent pattern from results over a long period are an invaluable guide. It is important, therefore, to ensure that methods are not changed without due consideration being given to the historical records of the plant. As a general principle, the bacterial counts tend to increase throughout plant runs, and it is important to know the pattern that these trends are likely to follow.

With the low $A_w$ of milk powder, one would expect the microbial content of the product to decrease with storage. Crossley (1962) confirmed that the population does decrease during storage, but that the rate varies with different powders. Mair-Waldburgh and Lübenau-Nestle (1974) observed decreases in microbial counts in spray- and roller-dried powders after storage at 20 °C for six months. Whilst a direct comparison was not possible, Mercurio and Tadjall (1979) observed that instant dried skim-milk stored in well-sealed tins for 20 years showed a microbial count comparable to that of a freshly-made product, a result that implies a fair degree of microbiological stability.

**Pathogenic Organisms**

The principal concern of microbiologists in relation to milk powder has been with the occurrence of pathogens. Specifically, *Salmonella* spp. and *Staphylococcus aureus* have been highlighted. As food poisoning micro-organisms, their presence in milk powder justifiably gives cause for concern and, furthermore, the wide and increasing use of milk powder as an ingredient of other foods makes the significance even greater.

Roller-dried milk has an exceptional record amongst foods for its freedom from pathogens.

Spray-dried milk, however, has come under the spotlight, and a number of outbreaks involving *Staph. aureus* have occurred in recent years. Hobbs (1955) referred to a number of outbreaks and suggested possible sources of contamination. Crossley and Campling (1957) investigated the process and concluded that a small proportion of *Staph. aureus* could survive spray drying. Crossley (1962) observed, however, that whilst the ubiquitous *Staph. aureus* occurred in raw milk supplies, there was no connection with the types which were isolated from the resultant milk powder. He concluded, therefore, that contamination within the plant environment was the most likely possibility. Miller (1972) and Heldman *et al.* (1968) identified a number of areas where there was contact between the air and the product, suggesting that these could provide access for *Salmonella* spp. However, if one accepts the argument of aerial contamination within the spray drying environment, then Miller's suggested points of access for *Salmonella* are applicable to any airborne contaminations.

In relation to laboratory methods, Keogh (1971) carried out the last comprehensive review of these available for the detection of *Staph. aureus*. Opinions differ on the efficacy of various methods, and the only reliable recourse is to the method laid down by the International Dairy Federation (IDF, 1978); this provides a reference method for the detection of coagulase-positive staphylococci in dried milk. However, Karolak *et al.* (1976) have compared a number of recent methods, and stated that the method of Baird-Parker (1962) gives the most precise results.

Alternative approaches to the enumeration of *Staph. aureus* have begun to emerge, and the determination of staphylococcal nuclease (Koupal and Diebel, 1978), and the use of radio-immunoassays for detecting staphylococcal enterotoxins (Miller *et al.*, 1978) look promising. The procedures have claimed detection of 10 ng and 1 ng of the nuclease and enterotoxin, respectively in 1 g quantities of product. The use of these new rapid procedures will no doubt increase in the course of time, and the longer, time-honoured method of colony production will be used for confirmation or reference.

The other principal pathogen which has occurred in dried milk is *Salmonella*. The occurrence of *Salmonella* in dried milks has been highlighted in the USA, and Schroeder (1967) after examining 3315 samples from 200 factories in 19 states, claimed that 1 % were contaminated with the organism. Collins *et al.* (1968) drew attention to the presence of *Salmonella newbrunswick* in instant milk, and suggested that the 'instantising' process was at fault.

The survival of *Salmonella* through the spray drying process has been a subject of investigation by a number of workers (McDonough and Hargrove, 1968; Licari and Potter, 1970). Miller and co-workers, using an experimental dryer, examined the effects of two different temperature conditions on the survival of *Salmonella* and *Escherichia coli*. They suggest that product temperature and particle density are major factors influencing survival; they also observed that a high fat content appeared to enhance the death of the micro-organisms concerned. There are risks in extrapolating data from small experimental dryers, due to the different residence times and heating patterns, to the situation found in the large commercial dryers. However, the data do form a guide, and at least endorse the need for constant vigilance and care. A point endorsed by Licari and Potter (1970) who showed that *Salmonella* spp. are not completely killed in spray drying. They used different inlet and outlet temperature patterns to Miller *et al*., and again the work, understandably, was carried out in a small laboratory dryer. Nevertheless, they were able to examine the effects of storage at different temperatures, and they reported that storage at 45 °C and 55 °C had a lethal effect on the test organisms—*S. typhimurium* and *S. thompson*. At temperatures of 25 °C and 35 °C, the numbers of *Salmonella* organisms were reduced in 4 to 8 weeks, but many were still present.

It could be concluded, therefore, that *Salmonella* spp. can survive the spray drying process even at the maximum commercial temperatures. However, this will be dependent on species, process conditions and the level of micro-organisms originally present. The conclusion for plant management must be a constant need for vigilance and care in plant hygiene and operation.

The vexing question of methodology in relation to *Salmonella* spp. has been extensively reviewed by Harvey and Price (1979). An important observation is that the isolation and identification of *Salmonella* spp. is of paramount importance. This confirms the widely held view amongst microbiologists that the isolation and identification of salmonellae is a procedure requiring adequate facilities and expertise, and that it should not be regarded as a routine procedure suitable for the control laboratories of small factories. Mossel and Shennan (1976) have emphasised the necessity for reliable and reproducible methods for product monitoring.

The question of using the coli–aerogenes group as an indicator for *Salmonella* spp. has also been reviewed by Mossel and Shennan (1976), who suggest that it is valid for dried milk due to the infrequent occurrence of lactase-negative enterobacteriaceae. The dairy industry is well attuned to the significance of the presumptive test for coli–aerogenes organisms, and whilst the evidence supporting its retention may be sparse, it is at least a precautionary procedure (Lück, 1980). Its recognition as an official IDF method reflects the commonality of opinion (IDF, 1974).

One major question in relation to the two principal pathogens is that of standards which are acceptable within the commercial sector. All too frequently specifications call for an organism to be absent without citing quantities or methods. It is important that specifications be written so as to be as precise and realistic as possible, and so avoid any danger of ambiguity. This is particularly important in relation to milk powders which are now an international commodity.

## REFERENCES

ADMI (1971) *Standards for Grades of Dry Milk including Methods of Analysis*, ADMI Bull. 91b, American Dry Milk Institute, Chicago, Ill.

BAIRD-PARKER, A. C. (1962) *J. Appl. Bact.*, **25**(1), 12–19.

BOCKELMANN, B. VON., (1969) *Milchwissenschaft*, **24**(8), 468–72.

CHOPIN, A., MOCQUOT, G. and LEGRAET, Y. (1977) *Can. J. Microbiol.*, **23**, 716–62.

CHOPIN, A., TESSONE, S., VILA, J-P. and LEGRAET, Y. (1978) *Can. J. Microbiol.*, **24**, 1371–80.

Code of Practice for manufacturers of dried milk (1954) Assoc. of British Manufacturers of Roller and Spray Process Milk Powders, London.

COLLINS, R. N. (1968) *J. Am. Med. Assn*, **203**(10), 838–44.

COX, W. A. (1970) *Chem. and Industry* (Feb), **7**, 223–9.

CROSSLEY, E. L. (1962) *Milk Hygiene*, WHO Monograph No. 48, Geneva.

CROSSLEY, E. L. and CAMPLING, M. (1957) *J. Appl. Bact.*, **20**(1), 65–70.

CROSSLEY, E. L. and JOHNSON, W. A. (1942) *J. Dairy Research*, **13**(1), 5–44.

GALESLOOT, Th. E. and STADHOUDERS, J. (1968) *Neth. Milk and Dairy J.*, **22**, 158–72.

GARRISON, E. R. (1946) *J. Bact.*, **52**, 45.

HALL, C. W. and HEDRICK, T. I. (1971) *Drying of Milk and Milk Products*, The AVI Publishing Company Inc., Westport, Connecticut.

HARVEY, R. W. S. and PRICE, T. H. S. (1979) *J. Appl. Bact.*, **46**(1), 27–56.

HAWLEY, H. B. and BENJAMIN, M. I. W. (1955) *J. Appl. Bact.*, **18**(3), 493–502.

HELDMAN, D. R., HALL, C. W. and HEDRICK, T. I. (1968) *J. Dairy Sci.*, **51**(3), 466–70.

HIGGINBOTTOM, C. (1944) *J. Dairy Res.*, **14**(1, 2), 184–94.

HOBBS, B. C. (1955) *J. Appl. Bact.*, **18**(3), 484–92.

IDF (1970) Standard method for determining the colony count of dried milk and whey powders (reference method), FIL–IDF, 49: 1970.

IDF (1974) Milk and milk products—count of coliform bacteria, FIL–IDF, 73:1974.

IDF (1978) Detection of coagulase +ve staphylococci in dried milk (reference method), FIL–IDF, 60A:1978.

KAROLAK, K., KARNICKA, H. and KNAUT, T. (1976) *Roczniki Instytutu Przemsyslu Mleczarskiego*, **18**(2), 47–57.

KEOGH, B. P. (1966) *Fd Technol. Aust.*, **18**(3), 126–33.

KEOGH, B. P. (1965) *Aust. J. Soc. Dairy Technol. Spray Drying of Milk*, Tech. Pub. No. 16.

KEOGH, B. P. (1971) *J. Dairy Res.*, **38**(1), 91–111.
KOUPAL, A. and DIEBEL, R. H. (1978) *Appl. and Environmental Microbiol.*, **35**(6), 1193–7.
LICARI, J. J. and POTTER, N. N. (1970) *J. Dairy Sci.*, **53**(7), Part I, 865–70; Part II, 871–76; Part III, 877–82.
LÜBENAU-NESTLE, R. and MAIR-WALDBURGH, H. (1978) *XXth Internat. Dairy Congress*, **E**.
LOVELL, H. R. (1972) *J. Soc. Dairy Technol.*, **25**(3).
LÜCK, H. (1980) In: *Dairy microbiology*, Vol. 2, Robinson, R. K. (Ed.), Applied Science Publishers Ltd, London.
MAIR-WALDBURGH, H. and LÜBENAU-NESTLE, R. (1974) *XIXth Internat. Dairy Congress*, **1E**, 553–4.
MATTICK, A. T. R., HISCOX, E. R., CROSSLEY, E. L., LEA, C. H., FINDLAY, J. D., SMITH, J. A. B., THOMPSON, S. Y., KON, S. K. and EDGELL, J. W. (1945) *J. Dairy Research*, **14**(1, 2), 116–59.
McDONOUGH, F. E. and HARGROVE, R. E. (1968) *J. Dairy Sci.*, **51** (10), 1587–91.
MERCURIO, K. C. and TADJALL, V. A. (1979) *J. Dairy Sci.*, **62**(4), 633–6.
MILLER, B. A., REISER, R. F. and BERGDALL, M. S. (1978) *Appl. and Environmental Microbiol.*, **36**(3), 421–6.
MOSSEL, D. A. A. and SHENNAN, J. L. (1976) *J. Fd Technol.*, **11**(3), 205–20.
MURRAY, J. G. (1975) *IFST Proc.*, **8**(2), 81–7.
SCHROEDER, S. A. (1967) *J. Milk and Fd Technol.*, **30**, 376.
TROUT, G. M. (1950) *Homogenized Milk*, Michigan State University Press, Michigan.
WAITE, R. (1964) *J. Soc. Dairy Technol.*, **17**(3), 122–31.
WALLGREN, K. (1964) *Svenska Mejeritdn*, **56**(2), 699–700.

# 7

# The Microbiology of Concentrated Milks

F. E. NELSON
*Emeritus Professor of Food Science,*
*Department of Dairy and Food Science,*
*University of Arizona, Tucson, USA*

Dairy products of reduced moisture content may be produced because of savings in transportation and merchandising costs related to the reduced volume and weight, and these products, with their greater concentration of milk solids, are useful in the manufacture of ice cream, candies and a variety of other food items. In some instances, desirable special properties result from one or more of the processing operations.

In this chapter the concern will be with the microbiology of three principal groups of products: (1) condensed and evaporated milks; (2) sweetened condensed milk; and (3) a miscellaneous group made up of condensed sour products, condensed whey and the 'retentates' obtained by reverse osmosis and ultrafiltration. Some of these products are little different from pasteurised milk from a microbiological standpoint, but heat, at the level of commercial sterilisation, confers excellent microbiological keeping quality on 'canned' evaporated milk. Sugar at a level inhibitory to most microbial growth retards spoilage in sweetened condensed milk, while the low pH of the condensed sour products both inhibits microbial development, and increases the microbiocidal effectiveness of heat.

## CONDENSED AND EVAPORATED MILKS

### Concentrated Milk

This name is commonly used for a condensed milk prepared for human consumption as fluid milk after appropriate dilution, but without further processing. It is ordinarily prepared from Grade A milk, usually with a 3:1 concentration. The processing is done under conditions that satisfy Grade

A standards, including a final pasteurisation. One of the advantages is the reduced space required in the refrigerator, but this can be offset by the need to dilute before use and, in addition, the keeping quality needs to be superior if advantage is taken of less frequent purchase. In some instances, the savings in terms of the reduced number of packages per unit volume of milk (hence easier storage and transport) have not compensated for the increased costs of processing, and the demands of route personnel that they be paid the same for delivering a quart of concentrate as for three quarts of milk. While considerable scientific and commercial interest was shown in this product in the 1950s, it has not become an important dairy product.

The raw milk supply for the concentrate is given a heat treatment approximating pasteurisation, and is then concentrated at minimum temperature to minimise changes in flavour or physical characteristics. The concentrate is then standardised, homogenised and pasteurised before packaging. Pasteurisation is usually at a somewhat elevated temperature, such as 79·4 °C (175 °F) for 25 s, as this compensates for the slight protective effect of the greater solids concentration and, along with the removal during concentration of any volatile off-flavours, is considered by some to impart a richness of flavour. The product has essentially the same keeping quality as pasteurised milk, the increased solids level not being great enough to inhibit microbial growth. Storage at 10 °C (50 °F) will permit growth of not only many of the thermoduric bacteria, but also of many post-pasteurisation contaminants, if such contamination has occurred, and the defects encountered are essentially the same as those in pasteurised milk similarly contaminated.

## Bulk Condensed Milk

This product is usually made from manufacturing-grade raw milk, and is used as a source of milk solids for candy, bakery products, ice cream and a number of other manufactured foods. Some may be made from Grade A raw milk and can be used in the standardisation of market milk. This latter product must be processed and handled according to the provisions of the recommended sanitation ordinance for condensed and dry milk products, and condensed and dry whey used in Grade A pasteurised milk products (USFDA, 1978). Condensing is usually in the range of 2·5:1 to 4:1, depending upon the use for which it is being prepared. The product is not sterilised during or after processing, so it contains a number of viable micro-organisms, and the concentration of milk solids is not great enough to inhibit microbial development. Keeping quality is limited, particularly when great care is not used to control post-heating contamination.

The methylene blue reduction test, with a reduction time of not less than 2·5 h, has been widely used for evaluation of the raw milk used but with better cooling on the farm, the shortcomings of this procedure for detection of psychrotrophic bacteria limit its usefulness; other microbial tests of comparable severity may be used. Rejection of milk that is obviously 'off' in flavour or in aroma should be done on the receiving platform. Considerable emphasis has been given to sediment testing, but the results are not directly related to the bacteriological quality of the milk. If the milk is to be held for any period prior to processing, it should be cooled to 4·6 °C (40 °F) or below to avoid a potential build-up of the microbial population, and the attendant chance that acidity or other microbially produced defects could become a problem.

If a skim-milk condensed product is to be made, the milk is preheated and separated before further processing. In making concentrated whole milk, the product is customarily homogenised. Standardisation for the desired fat:solids ratio may precede condensing, or may be done at the same time as the product is standardised for total solids content following condensing. Prior to condensing, the milk is heated in a continuous preheater, or in a 'hot well', and heating is commonly to 65·6–76·7 °C (150–170 °F). This preheating temperature may be increased to 82·2–93·3 °C (180–200 °F) for as much as 15 min to increase viscosity and to impart other desirable characteristics to the product for use in special applications. The preheating is usually not controlled to the same degree as pasteurisation would be, but the microbiological results are much the same when temperatures in the upper part of the range are employed. However, the product should not be labelled 'pasteurised' because of the lack of proper safeguards. In the lower range of preheating temperatures, the product is usually not held at the forewarming temperature for any significant period of time, particularly if the process is continuous, and thus the microbiological destruction would be less than with pasteurisation. The heater, and the 'hot wells' or surge tanks, can serve as incubators for thermophilic bacteria under these latter conditions, and when such equipment is operated for long periods without intermediate clean-up, or when the milk supply contains excessive numbers of thermophilic bacteria, the numbers may build up to a point where acid and unclean flavours result.

The preheated milk is then concentrated in a vacuum pan, or in a multiple-effect evaporator. The exact temperatures and times at which it is processed depend greatly on the type of equipment employed, and the desired characteristics of the final product. A temperature range of 54·4–57·2 °C (130–135 °F) is not uncommon, and this is very suitable for the growth of thermophilic bacteria. Operation over an extended period provides an opportunity for development of a considerable thermophilic population, so that proper sanitation and control in the preceding phases of the operation become essential.

The heat treatments prior to, and during, condensing are appreciably less than adequate to provide a sterile product. The cooling, standardisation and packaging operations provide opportunities for contamination, particularly from improperly cleaned and microbiocidally-treated equipment and containers. In addition, the material(s) used for standardisation must be of good microbiological quality, for one must emphasise that heat treatments, effective as they can be on micro-organisms present at the time of treatment, leave no residual microbiocidal effect to act upon subsequent contaminants, or to limit the growth of survivors of the heat treatment. Also, the solute concentration in the condensed product is not sufficient to inhibit the growth of micro-organisms. Since the product does contain micro-organisms that have survived processing, and usually contains micro-organisms that have contaminated the product subsequent to heat treatment, keeping quality is limited. Quick cooling to 4·2 °C (40 °F) or below is essential. Even at this temperature, keeping quality may be limited if post-heating contamination has been high and, at higher temperatures, the time over which the product can be held without serious microbial spoilage is frequently very limited.

**Evaporated Milk**

This product has much in common with bulk condensed milk through the condensing stage. However, it is given long-term keeping quality at room temperature by commercial sterilisation, either before or after placing in a hermetically sealed container to protect against subsequent contamination. Most of the product is made from whole milk. The Federal Standards require at least 7·9 % milk fat and 25·9 % total solids in the final product, and the addition of limited amounts of salts for stabilisation is permitted. Some product of this type is made from skim-milk, and limited amounts are made with other fats substituted for milk fat, the resulting product being known as a 'filled' milk. The Evaporated Milk Association, a trade group of manufacturers of the product, has established sanitary and other standards. These standards involve microbiological and sediment tests for the incoming milk, as well as procedures for the inspection of producing farms and their operations; some aspects of the processing plant and its operation also are specified. The methylene blue reduction test, with a requirement that reduction does not occur in 2·5 h or less, is the most widely used microbiological test on raw milk, but resazurin reduction tests and direct microscopic counts are used by some.

The milk is frequently clarified centrifugally. This process removes somatic cells (leucocytes) and some bacteria, but the change in bacterial count is not significant. The materials used to standardise the fat:solids ratio, whether cream or skim-milk should be of the same or better quality than the raw milk, and this means that proper provisions must be made for the production and handling of these products.

Stability of the milk to the sterilising temperatures used later is essential, and the milk proteins are ordinarily stabilised against subsequent coagulation by heating to above 93·4 °C (200 °F) and holding for some minutes. This may be done in a 'hot well' with holding for as long as 20–25 min, but in larger continuous operations, the enclosed system may permit as high as 121 °C (250 °F) for a few minutes. Any of these temperatures destroys all of the non-sporulating bacteria present, and many of the less-resistant sporulating types; any bacteria that might cause infectious disease would also be killed. However, the enterotoxin produced by enterotoxigenic staphylococci would not be inactivated, making it essential that proper precautions are taken to prevent extensive growth of these organisms in the incoming milk. The temperatures employed are too high to permit the growth of thermophilic bacteria, so these organisms are not a problem.

In the actual condensing operation, whether it be batch or continuous, temperatures seldom exceed 54·5 °C (130 °F), and may be significantly lower under some circumstances; these temperatures will not kill those bacteria which may have survived preheating. Thermophilic bacteria may grow under these conditions, and may become a factor in limiting the length of time the equipment can be operated before a shut-down for cleaning becomes necessary. Operation for extended periods may also result in such extensive build-up of solids on the heating surfaces that satisfactory cleaning will be made quite difficult.

Following condensing, the product is homogenised, and the usual precautions concerning homogeniser care must be observed to avoid excessive contamination from this source. The product is then cooled and placed in storage, where the final standardisation of composition takes place. Holding under good refrigeration until packaging and sterilisation take place is essential, because the product is not sterile, and can thus spoil if conditions permit appreciable microbial growth.

The old procedure, of placing the evaporated milk in cans, sealing hermetically, and then sterilising by heat has been replaced in many operations by a system for sterilising the milk first, and then aseptically packaging it. The fillers used for placing the milk in the final container are complex, difficult to clean properly, and quite demanding in the procedures necessary for adequate microbiocidal treatment; one recommended procedure calls for, essentially, complete disassembly of the canner, and for careful cleaning of each of the component parts. In the absence of sound practice, the filler can contribute enough bacteria of the wrong types to require an increase in severity of the heat treatment necessary for 'commercial sterilisation' of the final product.

The cans are usually fabricated in an adjacent plant, and the heat used in fabrication is enough to ensure that the can will contribute few, if any, bacteria to the canned product. Because the milk is cold when placed in the can, sufficient head-space must be allowed for the expansion that occurs during the sterilisation process, otherwise, the internal pressure may open one or more of the seams, and the resulting 'leak' will permit the entry of bacteria when the heat level is no longer adequate for their destruction. The container closure must be hermetic, that is, must not permit the passage of air or fluid in either direction, and a non-hermetic seal is an invitation to contamination. Leak detection is usually accomplished by a device that checks for cans that, both before and after sterilisation, are beyond a range which has been established as 'normal'. Automatic weighing may be used to check the amount of fill, discarding cans which are beyond the normal range, and weighing will also detect spillage before, during or after closure, or through leaks during processing and cooling. With the development of autoclavable pouches, and appropriate methods of sealing glass containers, more use of these types of containers may occur in the future.

To have adequate keeping quality at room temperature, evaporated milk must be 'commercially sterile'. This means that it must not contain organisms which will grow, and probably produce defects, under the normal storage conditions. If a residual organism is prevented from growing by the lack of oxygen in the environment (a highly aerobic organism), or if it is an obligately thermophilic organism which will grow only at elevated temperatures such as 45 °C (113 °F), the product may be 'commercially sterile'. By appropriate laboratory procedures, either of these types could be recovered, even though they could not grow in the evaporated milk. The presence of thermophilic bacteria which survive the usual heat treatment can become a problem when the product is held at unusually warm temperatures, such as in a poorly ventilated warehouse in a tropical climate. Product going to such an environment may need to be processed at a slightly elevated temperature to provide the necessary keeping quality.

No operative problem would be involved in treating evaporated milk with a time–temperature combination that would provide absolute sterility, but such absolute sterility would be associated with an unacceptable level of 'cooked' flavour, a dark colour, and probably some modified physical characteristics. Therefore, the heat exposure customarily chosen is the minimum one that will provide 'commercial sterility' under that particular set of conditions, thus keeping the modifications of flavour and physical characteristics to a minimum. However, as Galesloot (1962) points out, an increase in temperature from 120° to 150°C will increase the relative spore destruction rate about 1000-fold, while the relative rate of browning will be increased only about 15·7-fold; with the markedly reduced time of exposure required at the higher temperatures to give an equivalent bacteriocidal effect, the retention of flavour and colour can be impressive. The relatively greater physico-chemical stability of the product, when higher temperatures combined with shorter times are used, results in a product of reduced viscosity (in some instances, to a degree that permits some fat separation) although, at the same time, shelf stability may be reduced by the tendency for gelation to occur. However, holding the product in the range of $100 \pm 15$°C for periods of time determined by appropriate tests can be employed to overcome any problems of low viscosity and gelation.

The historical procedure for sterilisation has been to hold the canned product for 15 to 20 min at 115°C (239°F), or slightly higher, and both the come-up and the cooling periods contributed some microbiocidal action, as well as affecting the flavour and physical properties. Both batch and continuous sterilisers have been used, the former usually only in smaller operations.

In the batch steriliser, coming-up, holding and cooling all take place in, what is, in effect, a large autoclave. To achieve even heating and cooling, the frame in which the cans are placed rotates around its longitudinal axis, providing movement of product within the cans to increase the rate of heat penetration. The steam is introduced over the whole length of the chamber through holes in a horizontal tube. During cooling, water is sprayed on the cans from a tube in the top of the autoclave, and care must be exercised to maintain a pressure balance to avoid bursting of the can seams.

The continuous steriliser customarily consists of three sections, preheater, steriliser and cooler, each with appropriate pressure-holding valves. In each section is a rotating reel which moves the cans spirally from entrance to discharge end, each can in its own compartment; the rolling of the cans accelerates the rate of heat exchange and encourages uniformity of heating. The tendency has been to increase the temperature and reduce the time of holding to gain some of the advantages of ultra-high temperature

(UHT) treatment, while still avoiding the problems of aseptic packaging.

As the technology of the process has improved, interest in the UHT heat treatment has increased, particularly when used in combination with aseptic packaging. The heating can be with tubular or plate heat exchangers using steam as the indirect heating medium, or by the direct injection of steam into the product, followed by a vacuum treatment to remove the water added by the steam. The use of falling films of product, or its dispersal as droplets have been two of the methods employed to permit very quick heating. The temperatures which are effective can only be achieved with both steam and product under pressure, so that the closed system must be engineered to withstand the necessary pressures safely; treatments from 130 °C (266 °F) for 30 s to 150 °C (302 °F) for less than 1 s have been employed. The time–temperature combination selected must be such that it will effectively kill a 'normal' load of mesophilic spore-forming bacteria, and *Bacillus subtilis* is probably the most resistant of this group. The highly heat-resistant, obligately thermophilic representatives of the species *Bacillus stearothermophilus* might survive such a treatment, probably in small numbers, but would not be expected to grow or be responsible for defects under the normal holding conditions for canned evaporated milk. When the product is to be held at higher temperatures, a treatment more stringent than usual might need to be employed and, similarly, when the load of any type of spore is excessive. Spores which develop at the optimum growth temperature for the organism have been reported to be the most resistant (Theophilus and Hammer, 1938), but the forewarming temperatures apparently constitute a sub-lethal heat shock which makes the spores more sensitive to the final heat treatment (Curran and Evans, 1945). For a more detailed discussion of factors which influence spore resistance, one should consult a treatise such as that by Stumbo (1972).

Aseptic packaging of UHT products poses numerous microbiological problems. The container and closure used must be sterile, as must the equipment through which the product passes, and contamination by micro-organisms from the air must be avoided. The container must be hermetically sealed, so that air, water and other sources of contaminants cannot gain access to the product. Avoidance of rough handling subsequent to filling is essential, as damage to the package or even temporary weakening of the closure may permit contamination. When metal cans are used, they may be flame-heated or autoclaved with superheated steam to be made sterile. Glass containers must normally be cleaned very thoroughly, and then autoclaved. When composite paper–plastic–foil containers are used, the blanks are customarily treated just before forming and filling under aseptic conditions, by a combination of hydrogen peroxide and dry heat; the air used for the latter purpose being heated to about 200 °C (392 °F) to effectively remove any residual peroxide.

All equipment and enclosures must be treated to destroy contaminating micro-organisms prior to use. Superheated steam is the usual treating agent, although hot air could be used for some parts under appropriate circumstances. One of the problems is that in the case of jamming or other malfunction that requires entry to the system, the sterilisation procedure must be repeated. An atmosphere of superheated steam or hot gas must be maintained around the filler, closing machine and interconnecting conveyor system to preserve sterility.

**Microbiological Examination**

Evaporated milk will usually show no viable organisms when examined by customary procedures immediately after packaging, and micro-organisms seldom develop even after prolonged holding at usual room temperatures, although defects do appear occasionally. In past years, much canned condensed milk was held for 2–3 weeks in a warm room to detect spoilage before shipping but, with improved technology and laboratory control, holding of the entire lot is seldom done. Representative cans will be held out for incubation and examination. Where the product is to be shipped to warmer areas, incubation of the cans at 37 or 55 °C is frequently used to detect facultative or obligate, respectively, thermophilic bacteria which may have survived the heat treatment. Small amounts of the package contents are removed, with great attention to completely aseptic conditions, smeared onto plates already poured with appropriate agar, and incubated at the temperature at which the packages had been held. Attempts at quantitative results are seldom made, since the presence of a viable organism is the criterion employed. An examination of colony type frequently will permit a decision to be made as to whether the organism(s) survived heat treatment, were contaminants during aseptic packaging, or gained entry as a result of a leaky container. The presence of non-spore-forming organisms indicates post-sterilisation contamination, frequently as the result of a leaky container, but the presence of spore-formers is usually associated with inadequate heat treatment of the product.

**Defects**

Many of the defects reported in the literature and summarised by Hammer and Babel (1957) are almost never found at the present time because of improvements in process technology and laboratory control by the industry. Microbial defects can be divided into those which are due to organisms of high heat resistance that survive a slightly inadequate heat treatment, and those which gain entrance after heat treatment, and which are usually of low heat resistance.

Most of the heat resistant forms are species of the genus *Bacillus*, although an occasional species of the genus *Clostridium* has been encountered. *Bacillus coagulans* and *B. stearothermophilus* may cause an acid coagulation and a slight cheesy odour and flavour. These organisms grow best at 37 °C (98·6 °F) and above, and high storage temperatures and/or inadequate cooling are factors in this type of spoilage. *Bacillus subtilis* causes a non-acid curd, which may then be digested to a brownish liquid with a bitter taste. The coagulum formed by *B. megaterium* is accompanied by some gas and a cheesy odour. Gas production associated with putrefaction and a smell of hydrogen sulphide have been reported as caused by a *Clostridium* sp., but this type of defect is very rare, Bulged cans are caused much more commonly by a chemical action on the metal of the can, or by overfilling of the cans with cold milk which then expands on heating.

Contamination subsequent to heating may result in a greater variety of defects. When a breakdown in control of the aseptic canning process occurs, the contaminating bacteria may be of both spore-forming and non-spore-forming types. When the defect is due to a leaky can the non-spore-forming types are usually the cause because no question of heat resistance is involved. Leaks in the hermetic seal of the container may be due to improper closure, subsequent corrosion or mechanical injury during subsequent handling; even a momentary leak may permit microbial entry.

Milk, with defects presumed to be caused by bacteria, does not always yield viable micro-organisms, and this may be due to misdiagnosis of what was actually a non-microbial defect. In other instances, the organisms may have grown and then died off before the product was examined, or the procedures used for isolation may not have been adequate for the particular organism involved. This is especially probable when a period of time has elapsed since the organism was growing, since long exposure to the adverse environment created by the metabolic products of growth may make the conditions necessary for recovery more exacting.

The consumer must be made aware of the perishability of evaporated milk once the can is opened. The heat treatment used to kill the organisms and provide keeping quality in the unopened can has no residual effect that will control the growth of subsequent contaminants, and opening the can under kitchen conditions is almost certain to lead to some contamination.

Careful cleaning of the can top and immersion in boiling water for several minutes will kill most, but not necessarily all, of the organisms on the can surface. Immersion of the opener in boiling water will reduce, but not entirely eliminate, contamination from this source. Any water added to dilute the milk for use, any utensil with which the milk comes in contact, and even the air are all apt to provide some microbial contamination. The product will keep for several days under refrigeration, if reasonable care is used to minimise contamination, but holding without refrigeration is almost certain to permit spoilage in as little as 24 h.

## SWEETENED CONDENSED MILKS

Sweetened condensed milk may be made either from whole milk or from skim-milk. The product made from whole milk must have at least 8 % fat and 28 % milk solids, while the standard for the type made from skim-milk is 24 % milk solids and less than 0·5 % milk fat. Water is evaporated and sugar added to yield a product with a sufficiently high solute concentration to prevent the growth of most micro-organisms. For the retail trade sweetened condensed milk is packaged in hermetically sealed, metal containers but for industrial purposes it is packaged in milk cans, barrels, steel drums and bulk tanks. Bakers, confectioners and the prepared food industry use considerable amounts of these products.

The keeping quality of sweetened condensed milk is largely the result of the increase in osmotic pressure (reduction in water activity), and the 'binding' of water by the added sugar. The increased concentration of milk solids brought about by the removal of water by evaporation also contributes to the increase in osmotic pressure, but this is relatively minor compared to the effect of added sugar. The absence of significant amounts of air in the hermetically sealed package also contributes considerably to the keeping quality of the canned product, inhibiting the growth of a number of aerobic micro-organisms, particularly moulds, a few yeasts and some micrococci, which can tolerate the high osmotic pressure.

The concentration of sugar-in-water of sweetened condensed milk is known as the 'sugar ratio', and it is calculated from the following formula.

$$\text{Sugar ratio} = \frac{\%\ \text{sugar in condensed milk}}{100 - \text{total milk solids in condensed milk}} \times 100$$

The ratio usually is 63·5–64·5 for the canned product, and about 42 for the bulk, whole milk product. The lower ratio for the latter is permissible because the storage is commonly fairly short, and refrigeration is used. The use of sugar to extend the shelf-life should not be considered a substitute for good quality raw milk, proper sanitation, or adequate processing and holding practices. The heat treatments employed in processing are insufficient to 'sterilise' the product, so residual organisms are always present to cause problems if the product is not handled satisfactorily.

The raw milk used in production usually is of 'manufacturing grade'. Among the bacterial standards suggested have been a methylene blue reduction time not exceeding 3·5 h, a resazurin reduction to not less than Munsell PBP 7/5·5 in the 1 h test, or a direct microscopic clump count not to exceed $10^6$ ml$^{-1}$. No physical abnormalities, such as flakes, clots, blood or insect parts should be apparent, and no acid development should have occurred; although some older standards did permit slight development. Particularly in some European areas, a stability test, i.e. mixing equal parts of milk and 75–80% alcohol should not give rise to curdling, has been used, even though results of the test do not always parallel the stability of the sweetened condensed milk.

The processing of sweetened condensed milk can be divided into forewarming, superheating, sugar addition, condensing, cooling, forced crystallisation and packaging. The only truly microbiocidal heat used is during the forewarming and superheating (if used) phases, so these must be depended upon to destroy all pathogenic micro-organisms, and also most of the potential spoilage micro-organisms. The most common temperature and times are 82–100 °C (180–212 °F) for 10–30 min, with occasional use of higher temperatures and shorter times in an enclosed chamber. All but the most heat-resistant types of organism will be destroyed, including potential spoilage agents, and hence the presence of the latter in the end-product is almost invariably due to contamination subsequent to forewarming. The natural enzymes of milk are also inactivated, but the proteolytic and lipolytic enzymes resulting from excessive microbial growth may not be affected and may cause problems in the final product. While many of the safeguards incorporated into the pasteurisation of fluid milk are not apparent here, the higher temperatures usually employed provide a margin of safety. Leaky valves, dead ends and other deficiencies which might allow some product to escape full treatment must be avoided.

The sugar normally added is sucrose, but other sugars may be used, at least in part, for special purposes. The sugar is normally an unimportant source of micro-organisms, but under unfavourable conditions, it may be contaminated with mould spores, osmophilic yeasts or bacteria that will produce acid and gas. Sugar storage should be in a dry place, free from dust, insect and rodent contamination. The sugar may be added as such to the forewarmed milk prior to entry to the vacuum pan, or as an approximately 65% solution late in the condensing operation. Addition prior to forewarming does reduce the microbiocidal effectiveness of the heating, but addition to the forewarmed milk may contribute to age thickening, so this latter procedure is used primarily for bulk product to be used quickly.

The condensing operation is carried out in a vacuum pan at approximately 57·2 °C (135 °F), but the temperature may be permitted to drop to 48·9 °C (120 °F) late in the cycle. Because it follows earlier heating at a considerably higher temperature, condensing cannot be expected to significantly reduce the microbial population. The vacuum pan and its associated equipment must be cleaned very well, and the microbiocidal treatment carried out very thoroughly to avoid having the pan become a source of contamination. The sticky nature of the product increases the difficulty of cleaning, and 'cooked on' material may become a real problem. An acid detergent following the usual alkali detergent treatment has been suggested to help control formation of milkstone.

The forced crystallisation step consists of seeding the partially cooled milk with very fine lactose crystals, usually when the temperature is approximately 30 °C (86 °F), to induce the formation of numerous small lactose crystals, rather than fewer large ones during subsequent cooling. The added crystals are usually not heavily contaminated, but conversion of the lactose to the α-anhydride form by heating under vacuum to 93·3 °C (200 °F), followed by fine grinding and autoclaving in sealed cans at 130 °C (266 °F) for 1–2 h, has been suggested for sterilisation.

The cans for the retail trade (usually 12 oz. for the skimmed, and 14 oz. for the whole milk product), as well as the lids, should be microbiocidally treated by gas flames, superheated steam or ultra-violet radiation. The fillers are usually of the plunger type, equipped with a cut-off to prevent dripping, and these fillers are quite complex and difficult to clean adequately. They have been the source of heavy contamination and serious outbreaks of spoilage in the packaged product. They should be disassembled after each day's run, thoroughly washed, steamed and stored dry, or otherwise this equipment can be a major source of micrococci, yeasts and moulds in sweetened condensed milk. The cans must be filled as full as possible without causing later bulging, as a minimum of air space plus an hermetic seal will help restrict the growth of the aerobic organisms that may cause defects. Imperfect seals, or can damage subsequent to closure, may permit the entry not only of micro-organisms, but also of the air which is necessary for the growth of a number of spoilage organisms.

Bulk product may be stored and shipped in containers varying in size from 10 gal. cans to railroad tank cars. The industrial product is usually used within a few days, but if it is not used quickly, it should have something approaching the high sugar content of the retail canned product. Surface growth of aerobic organisms may be restricted by having the containers full, and ultra-violet lights over the storage tanks, along with protection from atmospheric contamination, may be used to combat surface mould. Provision must also be made to avoid any surface dilution, such as by condensate from above the liquid line.

**Microbiological Examination**

Viable micro-organisms are commonly found in the final product, and some reports place the numbers from a few hundred to 100 000 $g^{-1}$. The heat treatments used are not adequate to kill spore-forming bacteria, and further processing and handling usually contribute a variety of micro-organisms; the sugar levels employed permit some types to grow if other conditions are favourable. Enough oxygen may be present in the head-space of an incompletely filled, or poorly sealed, container to permit the growth of organisms able to tolerate the high osmotic pressure (reduced water activity) of the product.

Sampling requires considerable care. Cans must be thoroughly cleaned, the area of opening treated adequately with heat and/or microbiocidal chemicals, the opening instrument sterilised, and great care used in the withdrawal of the product. In bulk product, representative sampling of the viscous material is a problem as adequate mixing is difficult and, in some instances where aerobic micro-organisms are involved, a sample from the product surface may be more revealing than a mixed sample.

Three types of routine counts are made on sweetened condensed milk: (1) the standard plate count; (2) a coliform count; and (3) a yeast and mould count, and the procedures are outlined in standard methods for the examination of dairy products (APHA, 1978). For these determinations, a primary 1:10 gravimetric dilution is made, as the viscosity of the product precludes an accurate volumetric initial dilution, even when the product is cautiously prewarmed to 45 °C to reduce viscosity. If further dilutions are necessary, volumetric procedures are satisfactory. If the product is coagulated, the initial dilution may be made more advantageously using 1·25% sodium citrate in the dilution blank. Some have used agar containing 25–30% sugar to favour osmophilic organisms that may be present, but this is not a standard procedure. High standard plate counts and the presence of coliform bacteria or yeasts and moulds in recently processed product are considered indices of contamination following preheating. Some organisms, particularly some of the bacteria (including the coliforms), may die off with holding, but micrococci, yeasts and moulds may proliferate.

**Defects**

With current improved technology, defects in sweetened condensed milk are relatively uncommon, but three types of defect of microbiological origin are still of some concern. Gas formation may be caused by yeasts of the genus *Torulopsis*, although coliform bacteria have been implicated occasionally when the sugar ratio has been in the 40–45 range. Since neither of these organism types is resistant to forewarming temperatures, contamination during subsequent processing is indicated. The defect is more common during warmer months. The cans may be bulged and blown by carbon dioxide, or a mixture of carbon dioxide and hydrogen, depending upon the organism involved. The defect usually develops slowly because of the slow growth rate of the responsible organisms in the high sugar concentrations encountered.

Thickening is usually accompanied by some acidity and cheesy odours, and this defect is encountered primarily in the bulk product of lower sugar content. Many species of bacteria have been mentioned as responsible, with the micrococci and spore-formers being encountered most frequently. Lower storage temperatures and improved plant sanitation have been found helpful in combating this defect.

Small masses of mould mycelium and coagulated casein (buttons) usually coloured white to brown, may be found on the surface or in the sub-surface layers, especially of the canned product because this is held for longer times; a disagreeable taste is associated with the defect. Species of *Aspergillus* and *Penicillium* have been implicated. The moulds will grow until all the available oxygen in the head-space is exhausted, although the buttons may continue to increase in size because of continued enzymic activity. Poor plant sanitation is a major factor, because these moulds are not highly heat resistant and do not survive forewarming. Under-fill of the cans increases available oxygen and thus favours this defect, but storage below 16 °C (60 °F) may be helpful in reducing the incidence of faulty cans.

## CONCENTRATED SOUR MILK AND WHEY

These products are made to utilise materials that otherwise might be wasted, or that frequently present a disposal problem at some dairy plants.

Many of these products are of reduced importance now that the quality of manufacturing grade milk has improved appreciably, and as alternative means of utilising the basic materials have been developed. The concentrates have a high acidity (pH below 4·5), and are thus inhibitory to the growth of most bacteria and a few yeasts. An absence of air is used to inhibit the growth of moulds and some aerobic yeasts. The solids concentration is not high enough to be adequately inhibitory, but it will slow the development of some organisms. Pasteurisation, or at least forewarming prior to concentration, will destroy many organisms.

Concentrated sour skim-milk is made from product soured by *Lactobacillus bulgaricus*. The milk is usually heated to 77–82 °C (170–180 °F) in a continuous heater, or 65–71 °C (150–160 °F) for 30 min in a vat, to destroy competing organisms, as well as to give a smoother body to the final product. After cooling to 40–46 °C (105–115 °F), the milk culture of *L. bulgaricus* is added; a 2 % inoculum level being usual. When the acidity has reached 1·7 to 2·0 %, calculated as lactic acid, the curd is broken up and the material drawn into the vacuum pan. Concentration is usually to about 28 % solids and 6 % acid, as this acid level is needed for good keeping quality. Nevertheless, the product will mould if not protected from contamination in an air-tight container. Semi-solid buttermilk is prepared in much the same way, although it should be warmed to about 71 °C (160 °F) before condensing to give the desired physical characteristics. Whey may also be handled in much the same way, although as it has been concentrated without acidification, it has to be held under refrigeration.

## RETENTATES

Retentates are the materials produced when selective membranes in combination with pressure are used to concentrate desirable components. Reverse osmosis is the term applied to the process separating low molecular weight components from their solvents, usually water, and the membranes employed permit passage into the discarded portion of only low molecular weight materials, such as water and some salts. Ultrafiltration is essentially a sieving process in which molecules such as proteins are retained, while the membrane is permeable not only to water, but also to solutes such as sugars. By selection of appropriate structural materials and use of various preparation techniques, a range of membranes having considerable variation of characteristics has been prepared. Cellulose acetate and its derivatives have been used for preparation of membranes for both procedures. Polysulphone and several related polymers have been used for membranes for ultrafiltration.

Glover *et al.* (1978) have prepared an extensive review on the uses of reverse osmosis and ultrafiltration on dairy products. These processes have been used to concentrate whole milk, skim-milk and whey, and they are alleged to be cheaper than evaporation in the 2× to 4× range of concentration. The retentates have been used as concentrated sources of solids for such products as ice cream and yoghurt, where some have found a better texture and flavour than when powder or bulk condensed milk was used. Ultrafiltration has been used for making soft cheese, such as Camembert, by concentrating the skim-milk by as much as a factor of 6×, with resultant savings in rennet required, and an increase of significant magnitude in the yield of cheese. Concentration of whey proteins, without subjecting them to the partial denaturation associated with concentration in the vacuum pan, has been achieved and ultrafiltration has been suggested as a way to produce low-lactose or lactose-free milks for lactose-intolerant people.

Certain microbiological problems are associated with these processes. The retention time, and the temperature during the processing, must be controlled to limit microbial growth. The higher the temperature, within the normal range used to minimise denaturation, the shorter the time that can be tolerated without excessive microbial growth in the product; in a few instances, temperatures high enough to inhibit microbial growth have been used. Once concentration has been achieved, the product should be used or processed further with minimum delay since the concentrates are not sterile and heating, which would reduce microbial populations, would negate some of the process advantages. Freezing might be advantageous under some circumstances to permit holding.

The membranes, which are the heart of the processes, require special cleaning and microbiocidal treatment. The cellulose acetate membranes must be kept moist and, as they will not tolerate high temperatures, such as those above 50 °C (122 °F), these membranes cannot tolerate the methods customarily used for cleaning of dairy equipment. One solution is to remove them from the system and treat them separately, but backwashing is helpful if the mounting of the somewhat delicate membrane will support the necessary backpressure. Some workers have suggested that the detergents used should contain proteolytic enzymes that will assist in the removal of substances fouling the membrane surface. Iodophors are among the most effective microbiocidal agents in use, a level of 10 ppm having been recommended. Some of the polysulphone flat sheet membranes withstand a wide pH range and temperatures up to 100 °C (212 °F), and traditional dairy cleaning methods may be used on these. A microbiocidal treatment with 0·1 % hydrogen peroxide has been suggested as an alternative.

Greater use of reverse osmosis and ultrafiltration by the food industry in the future appears quite probable, and with further experience will come a better understanding of the microbiological considerations involved, permitting more specific recommendations to be made.

## REFERENCES

APHA (1978) *Standard Methods for the Examination of Dairy Products*, 14th edn., American Public Health Association, Washington, DC.

Curran, H. R. and Evans, F. R. (1945) *J. Bacteriol.*, **39**, 335–46.

Galesloot, T. E. (1962) The sterilization of milk. In: *Milk Hygiene*, World Health Organization, Geneva, Switzerland.

Glover, F. A., Skudder, P. J., Stothart, P. H. and Evans, E. W. (1978) *J. Dairy Research*, **45**, 291–318.

Hammer, B. W. and Babel, F. J. (1957) *Dairy Bacteriology*, 4th edn., John Wiley & Sons, Inc., New York, NY.

Stumbo, C. R. (1972) *Thermobacteriology in Food Processing*, 2nd edn., Academic Press, Inc., New York, N. Y.

Theophilus, D. R. and Hammer, B. W. (1938) Influence of growth temperature on the thermal resistance of some bacteria from evaporated milk, *Iowa Agric. Expt. Sta. Research Bull.*, 244.

USFDA (1978) *Recommended sanitation ordinance for condensed and dry milk products and condensed and dry whey used in Grade A pasteurised milk products*, Superintendent of Documents, US Government Printing Office, United States Food and Drug Administration, Washington, DC.

# Index